W9-CRB-812

Nucleic Acids: Curvature and Deformation

ACS SYMPOSIUM SERIES **884**

Nucleic Acids: Curvature and Deformation

Recent Advances and New Paradigms

Nancy C. Stellwagen, Editor
University of Iowa

Udayan Mohanty, Editor
Boston College

Sponsored by the
ACS Division of Physical Chemistry

American Chemical Society, Washington, DC

Library of Congress Cataloging-in-Publication Data

Nucleic acids: curvature and deformation : recent advances and new paradigms / Nancy C. Stellwagen, Udayan Mohanty, editors.

 p. cm.—(ACS symposium series ; 884)

 "Sponsored by the ACS Division of Physical Chemistry"

 Includes bibliographical references and index.

 ISBN 0–8412–3862–6

 1. Nucleic acids—Conformation—Congresses.

 I. Stellwagen, Nancy C. II. Mohanty, Udayan. III. American Chemical Society. Division of Physical Chemistry. IV. Series.

QP624.5.S78C87 2004
572.8—dc22   2004041122

The paper used in this publication meets the minimum requirements of American National Standard for Information Sciences—Permanence of Paper for Printed Library Materials, ANSI Z39.48–1984.

Copyright © 2004 American Chemical Society

Distributed by Oxford University Press

PRINTED IN THE UNITED STATES OF AMERICA

Foreword

The ACS Symposium Series was first published in 1974 to provide a mechanism for publishing symposia quickly in book form. The purpose of the series is to publish timely, comprehensive books developed from ACS sponsored symposia based on current scientific research. Occasionally, books are developed from symposia sponsored by other organizations when the topic is of keen interest to the chemistry audience.

Before agreeing to publish a book, the proposed table of contents is reviewed for appropriate and comprehensive coverage and for interest to the audience. Some papers may be excluded to better focus the book; others may be added to provide comprehensiveness. When appropriate, overview or introductory chapters are added. Drafts of chapters are peer-reviewed prior to final acceptance or rejection, and manuscripts are prepared in camera-ready format.

As a rule, only original research papers and original review papers are included in the volumes. Verbatim reproductions of previously published papers are not accepted.

ACS Books Department

Contents

Indexes

Preface

Recent advances in theory and experiment have suggested that new paradigms are needed to describe sequence-dependent curvature and deformation in nucleic acids and protein–nucleic acid complexes. Research in this field is escalating not only because of the wealth of sequence information generated by the human genome project but also because of recent improvements in the experimental and theoretical methods used to determine nucleic acid and nucleic acid–protein structures.

The results from a variety of studies have indicated that a combination of structural, energetic and thermodynamic data will be needed to understand the sequence-dependent structure of nucleic acids and the role played by the energy of deformation in determining the structure and specificity of protein–nucleic acid complexes. An American Chemical Society (ACS) symposium to explore these topics was held in New Orleans, Louisiana in March 2003. This book contains a representative sample of the papers that were presented at the symposium. All chapters were peer-reviewed by two independent, anonymous referees.

The 12 chapters in this book describe a variety of topics at the forefront of research into the sequence-dependent curvature and deformation of DNA, RNA, and protein–nucleic acid complexes. Chapters describing theory and experiment are intermingled to emphasize that both fields have contributed importantly to our current understanding of nucleic acid structure and deformation.

Publication of this volume is timely, because several years have passed since the last overview of the field was published ("Twenty Years of DNA Bending" by Wilma Olson and Victor Zhurkin in: *Biological Structure and Dynamics;* Sarma, R. H.; Sarma, M. H., Eds.; Adenine Press: Schenectady, NY, 1996). We envision that this book will be of interest to graduate students and postdoctoral fellows who are interested in the structural biology of nucleic acids and protein–nucleic acid complexes, as well as investigators who are actively working in the field. The extensive lists of references given in many of the chapters will make the book a useful resource for persons entering the field and for seminar-based graduate and undergraduate classes.

We thank the ACS and the Petroleum Research Fund for financial support of the symposium. Research support from the National Institute of General Medical Sciences (N.C.S.) and National Science Foundation (U.M.), which made possible the preparation of this volume, is also gratefully acknowledged.

Nancy C. Stellwagen

Department of Biochemistry
University of Iowa
Iowa City, IA 52242
319–335–7896 (telephone)
319–335–9570 (fax)
nancy-stellwagen@uiowa.edu (email)

Udayan Mohanty

Eugene F. Merkert Chemistry Center
Boston College
Chestnut Hill, MA 02467
617–552–3610 (telephone)
617–552–2705 (fax)
mohanty@bc.edu (email)

Nucleic Acids: Curvature and Deformation

Chapter 1

Curvature and Deformation of Nucleic Acids: Overview

Nancy C. Stellwagen[1] and Udayan Mohanty[2]

[1]Department of Biochemistry, University of Iowa, Iowa City, IA 52242
[2]Eugene F. Merkert Chemistry Center, Boston College,
Chestnut Hill, MA 02467

The Past

The structure of DNA proposed by Watson and Crick (*1*) from fiber diffraction data (*2*) was symmetric and highly regular, leading to the idea that DNA structure was monolithic and independent of sequence. Subsequent studies, however, especially by Arnott and coworkers (*3*), showed that DNA structure is in fact highly variable, and depends on hydration, base-pair composition and the identity of the counterion.

Many of the early investigations of DNA conformation in solution revolved around the questions of polymer physics—the size, shape, and flexibility of the macromolecule and its stability under a variety of experimental conditions (reviews: *4-7*). The modern era of DNA structural studies began when the x-ray crystal structure of a dodecamer containing the *Eco*RI recognition site was solved by Dickerson and coworkers (*8-10*). This structure showed that each base-pair step makes its own individual contribution to the conformation of the helix, indicating that DNA structure must be sequence-dependent. A host of other x-ray and NMR studies of small DNA oligomers and DNA-drug and DNA-protein complexes followed, providing atomic-level detail about DNA structure and conformation in a variety of contexts (reviews: *6, 7, 11-14*). These and other studies were made possible by concomitant advances in molecular biology and the development of methods for automated oligonucleotide synthesis, which made it possible to prepare the large quantities of highly purified DNA oligomers needed for biophysical studies.

The next step in understanding sequence-dependent DNA structure came from the discovery that bent or curved DNA molecules migrate anomalously slowly in polyacrylamide gels (*15, 16*). An explosion of studies then followed, especially from the Crothers, Hagerman and Diekmann laboratories (reviews: *17-20*), which led to the important discovery that DNA bending was often associated with the presence of A-tracts (runs of four or more adenine residues in a row) that occurred every ten base pairs, in phase with the helix repeat. These results reinforced a previous suggestion by Trifonov (*21*), that naturally occurring DNA molecules contain a structural code based on phased ApA base-pair steps.

The techniques of x-ray diffraction and polyacrylamide gel electrophoresis were soon augmented by other methods, such as ligase-catalyzed cyclization, pioneered by Shore and Baldwin (*22, 23*) and applied to DNA oligomers by Ulanovsky and coworkers (*24*), the circular permutation assay, which was developed by Wu and Crothers (*25*) to determine the precise location of a bend within a given DNA fragment, and fluorescence resonance energy transfer (FRET), which was used extensively by Lilley and coworkers (*26*) to determine the conformation of 3- and 4-way junctions. Less frequently, other experimental methods, such as transient electric birefringence (27-30), circular dichroism (*31, 32*), hydroxyl radical footprinting (*33*) and Raman spectroscopy (*34, 35*) were used to characterize bent or curved DNA molecules and compare their properties with those of normal, mixed-sequence DNAs.

The various methods of characterizing sequence-dependent DNA bending led to a fundamental dichotomy. The gel and solution studies suggested that oligomers containing phased A-tracts were bent toward the minor groove, with an average bend angle of 18° - 20° per A-tract (*36-39*). However, the x-ray diffraction studies indicated that A-tracts were straight, and bending occurred at the junctions between the A-tracts and the rest of the DNA molecule (*40-43*), in a direction perpendicular to that inferred from the gel and solution studies (*43, 44*). Based on the x-ray studies, it was therefore proposed that A-tracts were straight and bending occurred in the GC-rich sequences flanking the A-tracts (*43*). These and other early studies of DNA bending and curvature are summarized in the excellent review by Olson and Zhurkin (*7*).

The Present

Studies of sequence-dependent DNA curvature and deformation were stalled at this point for several years, in part because each of the methods used to characterize DNA bending has its own set of limitations. The polyacrylamide gel studies are only semi-quantitative, because no theory adequately relates anomalously slow electrophoretic mobilities to DNA bending and because the observed mobility anomalies are dependent on the gel running conditions (*20,*

45-48). On the other hand, x-ray crystallography cannot detect long-range curvature of the helix backbone; the x-ray results are also confounded by uncertainties associated with crystal packing effects *(49, 50)* and the dehydrating agents used to induce precipitation *(51-53)*.

Further progress in understanding DNA bending and curvature began in the mid-to-late 1990s, due in part to advances in theory and technology and in part to several clever experiments carried out by young investigators. The revolution was led by Maher and coworkers *(54, 55)*, who showed by gel electrophoretic methods that neutralization of the phosphate residues on one side of the DNA helix led to bending of the helix axis in the direction of the neutral patch, due to electrostatic repulsions between the negatively charged phosphate groups on the other side of the helix. This experiment, which was based on a previous suggestion by Manning and coworkers *(56)*, led to the proposal that asymmetric neutralization of the phosphate residues could be a general mechanism for DNA bending. Maher and coworkers discuss the implications of phosphate crowding and DNA bending in Chapter 5 of this volume. Mohanty and coworkers present a theoretical description of asymmetric charge neutralization in Chapter 6.

The second critical experiment, carried out by Williams and coworkers a few years later *(57-59)*, was the redetermination of the x-ray crystal structure of the *Eco*RI dodecamer using synchrotron radiation and cryogenic temperatures, increasing the resolution substantially. The results suggested that the minor groove of the dodecamer was partially occupied by monovalent counterions localized near the central ApT base pair step, in agreement with the idea that A-tract bending is an electrostatic effect due to the asymmetric neutralization of phosphate residues *(60, 61)*. This new x-ray structure led to a flurry of activity by other crystallographers applying modern methods to the determination of DNA oligomer structures. The results indicate that some, but not all, monovalent cations are located some of the time in the minor groove of A-tract-containing DNAs, as described by Egli and Tereshko in Chapter 4 of this volume.

Other experiments during this period also suggested that preferential counterion binding occurs in the minor groove of A-tract DNAs. NMR experiments using the ^{23}Na magnetic dispersion technique showed that Na$^+$ ions are preferentially localized in the minor groove of dodecamers containing central A_2T_2- and A_4T_4-tracts, with a site occupancy of 1-5% *(62)*. Other NMR experiments showed that NH_4^+ ions are located in the minor groove at the 3'-end of A-tracts *(63)*, while Mn^{++} ions in the minor groove are located at the 5'-end of A-tracts *(64)*. In agreement with these results, capillary electrophoresis experiments showed that double-stranded DNA oligomers containing phased A_n- or A_nT_n-tracts migrated more slowly in the electric field than oligomers of the same size containing phased CACA-tracts, suggesting that excess counterions were bound to the A-tract-containing oligomers, decreasing their effective charge (*65*).

At about the same time, it was discovered that weakly oriented matrices could be used to increase the number of long-range NMR coupling constants observed for large asymmetric macromolecules (*66*). MacDonald and coworkers (*67*) used this technology to show that a dodecamer containing a central A_6-tract was intrinsically bent by 19° in the direction inferred from earlier solution and gel electrophoresis experiments. A model describing the molecular basis of A-tract bending is presented by MacDonald and Lu in Chapter 3 of this volume.

These experimental advances were accompanied by significant progress in the theoretical description of sequence-dependent DNA conformation and structure. Improvements in the force fields used to describe DNA energetics, accompanied by increased computer power, have allowed molecular dynamics simulations to be carried out over realistic, biologically relevant time intervals. Studies by the Beveridge group (*68, 69*) predicted that monovalent counterions would be localized in the minor groove of A-tract DNAs, particularly at the ApT step, as observed experimentally. Additional results are described by Beveridge and coworkers in Chapter 2 of this volume, set in the context of an extensive review of the literature describing DNA bending.

The flexibility or deformability of the DNA helix is also an important determinant of DNA structure. DNA flexibility is usually described in terms of its persistence length, which has been measured by many different investigators over the years, using many different techniques. It has generally been found that the persistence length of DNA ranges from 50 - 60 nm in solutions of low ionic strength, and decreases to ~45 nm in solutions containing 100 mM Na^+ or low concentrations of Mg^{++} ions (*70-77*).

The binding of proteins and other ligands to DNA can lead to deformation of the helix axis, which has determined for many DNA-protein complexes by x-ray diffraction and NMR methods (reviews: *7-13, 78-82*). The availability of these detailed atomic structures, along with other experimental measures of sequence-dependent structural variability (*83-85*), has led to the creation of conformational maps (*81, 86-88*) and "bendability" indices (*89*) for the various DNA base pair steps.

The contribution of energetics, ion binding and hydration to the stability and folding of curved DNA molecules in the presence and absence of small ligands is described by Marky and coworkers in Chapter 10 of this volume. An improved method of measuring protein-induced DNA bending by the ligase-catalyzed cyclization of DNA fragments containing single-stranded regions is described by Vologodskii in Chapter 11. The role of DNA curvature and flexibility in determining the stability and organization of nucleosome is described by de Santis and coworkers in Chapter 12.

The linear, double-stranded structure of the DNA helix is significantly distorted by the formation of 3-way and 4-way junctions, which are important intermediates in homologous genetic recombination and DNA repair (*90*). The stability of the junction and the angle between the nucleotide arms depends on the sequence at the junction and the presence or absence of divalent cations in the medium (*91, 92*). The structural and dynamic properties of 4-way junctions are reviewed by Lilley in Chapter 7 of this volume.

Important advances have also been made in understanding RNA structure, due in part to the increasing availability of x-ray and NMR structures of small RNAs and RNA motifs (*93, 94*) and to the high resolution x-ray crystal structure of the ribosome (*95, 96*). RNA structure is inherently more variable than DNA, because RNAs contain Watson-Crick and non-Watson-Crick base pairs, hairpins and internal loops of various sizes, base triples, base quadruples and pseudoknots (*97*). In chapter 8, Greenbaum and Lambert describe the role of a conserved pseudouridine residue in determining the structure of a spliceosomal pre-mRNA branch site. Kelley and coworkers describe the structural perturbations in tRNAs induced by disease-related mutations in human transfer RNAs in Chapter 9.

The Future

The mechanism by which the RNA polynucleotide chain is folded into its final equilibrium structure remains an open question. RNA folding appears to be a hierarchical process, in which small "folding motifs" are formed rapidly, followed by the slower association of these motifs into larger subunits and, finally, the complete molecule. The identification of the basic folding motifs is still a matter of debate (*97, 98*). The role of metal ions in the folding process is also being elucidated, using standard technologies (reviews: *97, 99-102*) as well as single molecule stretching experiments (*103*). Because the folding process appears to be hierarchical, it seems possible that the "RNA folding problem" may be solved in the near future. However, it is not yet clear whether the folding pathways observed *in vitro* will be the same as observed *in vivo*, where folding is modulated by the presence of proteins and other constituents in the cell (*104, 105*). There is also a need for a better theoretical understanding of RNA structure and the energetics of various folding pathways.

Preferential counterion binding to A-tract DNAs remains a controversial topic, primarily because the limited residence time of the counterions in the minor groove makes the interpretation of the x-ray results difficult (*106-108*) and the effects measured by other techniques (*e.g.*, *62, 65, 66*) small. In addition, there is the "chicken and egg" question. If preferential monovalent counterion binding in the minor groove of A-tract DNAs occurs, does ion binding cause A-tract bending because of asymmetric charge neutralization and

6

electrostatic collapse, as proposed by some (*60, 61, 109, 110*)? Or, do intrinsically bent A-tract DNAs sequester monovalent ions in the minor groove because the groove width is ideal for ion binding, as proposed by others (*106, 111*)? The resolution of these questions awaits further experimental and theoretical developments.

Another proposal that awaits future evaluation is the hypothesis that DNA major and minor grooves are sequence-specific, flexible ionophores (*110*). Since GC-rich sequences preferentially localize cations in the major groove (*111-117*) and undergo the B- to A-form structural transition relatively easily (*118-122*), and A-tract DNAs preferentially bind cations in the minor groove (*57, 58, 107, 109, 123-125*) and have the relatively rigid B'-DNA structure (*40, 41*), Hud and coworkers (*110*) have proposed that DNA conformation is determined by the relative proportion of counterions preferentially localized in the major and minor grooves of GC-rich and A-tract sequences, respectively. While much experimental data can be correlated by this hypothesis, it is possible that the ionophore concept may be more applicable to DNA and RNA structures that require counterion binding for stability.

DNA hydration, and its relation to the bending and deformation of the helix axis, is also imperfectly understood. It is well known that DNA hydration is sequence dependent (*126, 127*), although it is not often remembered that A-form DNA is stable up to 98% relative humidity and B-form DNA is stable down to 66% relative humidity (*3*). The "spine of hydration" on the floor of the DNA minor groove (*9*) appears to be covered by one or more layers of relatively well-ordered water molecules (*57, 58, 106, 116, 124, 128, 129*). The transient localization of cations on the floor of the minor groove must disrupt this ordered water structure, although how it happens is not clear (*130*). It is possible that the counterions are incorporated into the hydration layer(s) within the minor groove and then diffuse along the groove for various distances before exchanging with cations in the bulk solvent (*131*). This type of mechanism would be consistent with the conformational flexibility of the phosphate residues observed in ultra-high resolution x-ray structures of DNA oligomers (*116, 128*) and with the Manning theory of counterion condensation (*132*).

Further progress in understanding the relationship between DNA sequence, hydration, counterion binding and curvature of the helix axis will undoubtedly require an extensive interplay between theory and experiment in the upcoming years. The tools appear to be in place to achieve rapid progress.

References

1. Watson, J. D.; Crick, F. H. C. *Nature* **1953**, *171*, 737-738.
2. Wilkins, M. H. F.; Stokes, A. R.; Wilson, H. R. *Nature* **1953**, *171*, 738-740.

3. Leslie, A. G. W.; Arnott, S.; Chandrasekaran, R.; Ratliff, R. L. *J. Mol. Biol.* **1980**, *143*, 49-72.

4. Bloomfield, V. A.; Crothers, D. M.; Tinoco, I., Jr. I. *Physical Chemistry of Nucleic Acids*, Harper & Row: New York, NY, 1974.

5. Bloomfield, V. A.; Crothers, D. M.; Tinoco, Jr., I. *Nucleic Acids: Structures, Properties, and Functions*, University Science Books: Sausalito, CA, 2000.

6. Saenger, W. *Principles of Nucleic Acid Structure*; Springer-Verlag: New York, NY, 1984.

7. Olson, W. K.; Zhurkin, V. B. In *Biological Structure and Dynamics*; Sarma, R. H.; Sarma, M. H., Eds.; Proceedings of the Ninth Conversation, State University of New York, Albany, NY 1995; Adenine Press: Schenectady, NY, 1996, pp. 341-370.

8. Drew, H. R.; Wing, R. M.; Takano, T.; Broka, C.; Tanaka, S.; Itakura, K.; Dickerson, R. E. *Proc. Natl. Acad. Sci. USA* **1981**, *78*, 2179-2183.

9. Drew, H. R.; Dickerson, R. E. *J. Mol. Biol.* **1981**, *151*, 535-556.

10. Dickerson, R. E.; Drew, H. R. *J. Mol. Biol.* **1981**, *149*, 761-786.

11. Steitz, T. A., *Structural Studies of Protein-Nucleic Acid Interaction*, University Press: Cambridge, 1993.

12. *RNA-Protein Interactions*; Nagai, K.; Mattaj, I. W., Eds.; Frontiers in Molecular Biology; IRL Press: Oxford, 1994.

13. *DNA-Protein: Structural Interactions*; Lilley, D. M. J., Ed.; Frontiers in Molecular Biology; IRL Press: Oxford, 1995.

14. Wahl, M. C.; Sundaralingam, M. *Current Biol.* **1995**, *5*, 282-295.

15. Marini, J. D.; Levene, S. D.; Crothers, D. M.; Englund, P. T. *Proc. Natl. Acad. Sci. USA* **1982**, *79*, 7664-7688; *op. cit. 80*, 7678 (correction).

16. Stellwagen, N. C. *Biochemistry* **1983**, *22*, 6186-6193.

17. Crothers, D. M.; Haran, T. E.; Nadeau, J. G. *J. Biol. Chem.* **1990**, *265*, 7093-7096.

18. Hagerman, P. J. *Annu. Rev. Biochem.* **1990**, *59*, 755-781.

19. Hagerman, P. J. *Biochim. Biophys. Acta* **1992**, *1131*, 125-132.

20. Diekmann, S. *Nucleic Acids Res.* **1987**, *15*, 247-265.

21. Trifonov, E. N. *CRC Crit. Rev. Biochem.* **1985**, *19*, 89-106.

22. Shore, D.; Baldwin, R. L. *J. Mol. Biol.* **1983**, *170*, 957-981.

23. Shore, D.; Baldwin, R. L. *J. Mol. Biol.* **1983**, *170*, 983-1007.

24. Ulanovsky, L.; Bodner, M.; Trifonov, E. N.; Choder, M. *Proc. Natl. Acad. Sci. USA* **1986**, *893*, 862-866.

25. Wu, H.-M.; Crothers, D. M. *Nature* **1984**, *308*, 509-513.

26. Lilley, D. M. J. *Q. Rev. Biophys.* **2000**, *33*, 109-159.

27. Hagerman, P. J. *Proc. Natl. Acad. Sci. USA* **1984**, *81*, 4632-4636.

28. Stellwagen, N. C. *Biopolymers* **1991**, *31*, 1651-1667.

29. Nickol, J.; Rau, D. C. *J. Mol. Biol.* **1992**, *228*, 1115-1123.

30. Chan, S. S.; Breslauer, K. J.; Austin, R. H.; Hogan, M. E. *Biochemistry* **1993**, *32*, 11776-11784.

8

31. Brahms, S.; Brahms, J. G. *Nucleic Acids Res.* **1990**, *18*, 1559-1564.
32. Herrera, J. E.; Chaires, J. B. *Biochemistry* **1989**, *28*, 1993-2000.
33. Burkhoff, A. M.; Tullius, T. D. *Cell* **1987**, *48*, 935-943.
34. Chan, S. S.; Austin, R. H.; Mukerji, I.; Spiro, T. G. *Biophys. J.* **1997**, *72*, 1512-1520.
35. Mukerji, I.; Williams, A. P. *Biochemistry* **2002**, *41*, 69-77.
36. Levene, S. D.; Wu, H.-M.; Crothers, D. M. *Biochemistry* **1986**, *25*, 3988-3995.
37. Koo, H.-S.; Drak, J.; Rice, J.A.; Crothers, D. M. *Biochemistry* **1990**, *29*, 4227-4234.
38. Dlakic, M.; Harrington, R. E. *J. Biol. Chem.* **1995**, *270*, 29945-29952.
39. Zahn, K.; Blattner, F. R. *Science* **1987**, *236*, 416-422.
40. Nelson, H. C. M.; Finch, J. T.; Luisi, B. F.; Klug, A. *Nature* **1987**, *330*, 221-226.
41. Coll, M.; Frederick, C. A.; Wang, A. H.-J.; Rich, A. *Proc. Natl. Acad. Sci. USA* **1987**, *84*, 8385-8389.
42. DeGabriele, A. D.; Sanderson, M. R.; Steitz, T. A. *Proc. Natl. Acad. Sci. USA* **1989**, *86*, 1816-1820.
43. Goodsell, D. S.; Kaczor-Grzeskowiak, M.; Dickerson, R. E. *J. Mol. Biol.* **1994**, *239*, 79-96.
44. Dickerson, R. E. *Nucleic Acids Res.* **1998**, *26*, 1906-1926.
45. Thompson, J. F.; Landy, A. *Nucleic Acids Res.* **1988**, *16*, 9687-9705.
46. Drak, J.; Crothers, D. M. *Proc. Natl. Acad. Sci. USA* **1991**, *88*, 3074-3078.
47. Ussery, D. W.; Higgins, C. F.; Bolshoy, A. *J. Biomolec. Struct. Dyn.* **1999**, *16*, 811-823.
48. Stellwagen, N. C. *Electrophoresis* **1997**, *18*, 34-44.
49. Dickerson, R. E.; Goodsell, D. S.; Neidle, S. *Proc. Natl. Acad. Sci. USA* **1994**, *91*, 3579-3583.
50. Shatzky-Schwartz, M.; Arbuckle, N. D.; Eisenstein, M.; Rabinovich, D.; Bareket-Samish, A.; Haran, T. E.; Luisi, B. F.; Shakked, Z. *J. Mol. Biol.* **1997**, *267*, 595-623.
51. Sprous, D.; Zacharias, W.; Wood, Z. A.; Harvey, S. C. *Nucleic Acids Res.* **1995**, *23*, 1816-1821.
52. Ganunis, R. M.; Guo, H.; Tullius, T. E. *Biochemistry* **1996**, *35*, 13729-13732.
53. Dickerson, R. E.; Goodsell, D.; Kopka, M. L. *J. Mol. Biol.* **1996**, *256*, 108-125.
54. Strauss, J. K.; Maher, L. J. *Science* **1994**, *266*, 1829-1834.
55. Strauss-Soukup, J. K.; Vaghefi, M. M.; Hogrefe, R. I.; Maher, L. J. *Biochemistry* **1997**, *36*, 8692-8698.
56. Manning, G.; Ebralidse, K. K., Mirzabekov, A. D.; Rich, A. *J Biomol. Struct. Dyn.* **1989**, *6*, 877-889.

57. Shui, X.; McFail-Isom, L.; Hu, G. G.; Williams, L D. *Biochemistry* **1998**, *37*, 8341-8355.
58. Shui, X.; Sines, C. C.; McFail-Isom, L.; VanDerveer, D.; Williams, L. D. *Biochemistry* **1998**, *37*, 16877-16887.
59. Sines, C. C.; McFail-Isom, L.; Howerton, S. B.; VanDerveer, D.; Williams, L. D. *J. Am. Chem. Soc.* **2000**, *122*, 11048-11056.
60. McFail-Isom, L.; Sines, C. C.; Williams, L. D. *Curr. Opin. Struct. Biol.* **1999**, *9*, 298-304.
61. Williams, L. D.; Maher, L. J. *Annu. Rev. Biophys. Biomol. Struct.* **2000**, *29*, 497-521.
62. Denisov, V. P.; Halle, B. *Proc. Natl. Acad. Sci. USA* **2000**, *97*, 629-633.
63. Hud, N. V.; Sklenář, V.; Feigon, J. *J. Mol. Biol.* **1999**, *286*, 651-660.
64. Hud, N. V.; Feigon, J. *Biochemistry* **2002**, *41*, 9900-9910.
65. Stellwagen, N. C.; Magnusdóttir, S.; Gelfi, C.; Righetti, P. G. *J. Mol. Biol.* **2001**, *305*, 1025-1033.
66. Bax, A.; Kontaxis, G.; Tjandra, J. *Methods Enzymol.* **2001**, *339*, 127-174.
67. MacDonald, D.; Herbert, K.; Zhang, X.; Polgruto, T.; Lu, P. *J. Mol. Biol.* **2001**, *306*, 1081-1098.
68. Young, M. A.; Jayaram, B.; Beveridge, D. L. *J. Am. Chem. Soc.* **1997**, *119*, 59-69.
69. Young, M. A.; Beveridge, D. L. *J. Mol. Biol.* **1998**, *281*, 675-667.
70. Hagerman, P. J. *Annu. Rev. Biophys. Biophys. Chem.* **1988**, *17*, 265-286.
71. Rizzo, V.; Schellman, J. *Biopolymers* **1981**, *20*, 2143-2163.
72. Kam, A.; Borochov, N.; Eisenberg, H. *Biopolymers* **1981**, *20*, 2671-2690.
73. Taylor, W. H.; Hagerman, P. J. *J. Mol. Biol.* **1990**, *212*, 363-376.
74. Sobel, E. S.; Harpst, J. A. *Biopolymers* **1991**, *31*, 1559-1564.
75. Smith, S. G.; Finzi, L.; Bustamante, C. *Science* **1992**, *258,* 1122-1126.
76. Bednar, J.; Furrer, P; Katritch, V.; Stasiak, A. Z.; Dubochet, J.; Stasiak, A. *J. Mol. Biol.* **1995**, *254*, 579-594.
77. Lu, Y.; Weers, B.; Stellwagen, N. C. *Biopolymers* **2002**, *61*, 261-275.
78. McGill, G.; Fisher, D. E. *Chem. Biol.* **1998**, *5*, R29-R38.
79. Pérez-Martín, J.; de Lorenzo, V. *Annu. Rev. Microbiol.* **1997**, *51*, 593-628.
80. Olson, W. K.; Gorin, A. A.; Lu, S.-J.; Hock, L. M.; Zhurkin, V. B. *Proc. Natl. Acad. Sci. USA* **1998**, *95*, 11163-11168.
81. Saeker. M.; Record, M. T., Jr. *Curr. Opin. Struct. Biol.* 2002, *12,* 311-319.
82. Kanhere, A.; Bansal, M. *Nucleic Acids Res.* **2003**, *31*, 2647-2658.
83. Harrington, R. E. *Electrophoresis* **1993**, *14*, 732-746.
84. Bolshoy, A.; McNamara, P.; Harrington, R. E.; Trifonov, E. N. *Proc. Natl. Acad. Sci. USA*, **1991**, *88*, 2312-2316.
85. Satchwell, S. C.; Drew, H. R.; Travers, A. A. *J. Mol. Biol.* **1986**, *191*, 659-675.
86. Packer, M. J.; Dauncey, M. P.; Hunter, C. A. *J. Mol. Biol.* **2000**, *295*, 71-83.

87. Packer, M. J.; Dauncey, M. P.; Hunter, C. A. *J. Mol. Biol.* **2000**, *295*, 85-103.
88. Gorin, A. A.; Zhurkin, V. B.; Olson, W. K. *J. Mol. Biol.* **1995**, *247*, 34-48.
89. Munteanu, M. G.; Vlahoviček, V.; Parthasarathy, S.; Simon, I.; Pongor, S. *TIBS* **1998**, *23*, 341-347.
90. Holliday, R. *Genet. Res.* **1964**, *5*, 282-304.
91. Lilley, D. M. J. *Q. Rev. Biophys.* **2000**, *33*, 109-150.
92. Thorpe, J. H.; Gale, B. C.; Teixeira, S. C. M.; Cardin, C. J. *J. Mol. Biol.* **2003**, *327*, 97-109.
93. Moore, P. B. *Annu. Rev. Biochem.* **1999**, *68*, 287-300.
94. Doudna, J.; Cech, T. *Nature* **2002**, *418*, 222-228.
95. Wimberly, B. T.; Brodersen, D. E.; Clemons, W. M., Jr.; Morgan-Warren, R. J.; Carter, A. P.; Vonhein, C.; Hartsch, T.; Ramakrishnan, V. *Nature* **2000**, *407*, 327-339.
96. Ban, N.; Nissen, P.; Hansen, J.; Moore, P. B.; Steitz, T. A. *Science* **2000**, *289*, 905-929.
97. Leontis, N. B.; Westhof, E. *Curr. Opin. Struct. Biol.* **2003**, *13*, 300-308.
98. Gutell, R. R.; Lee, J. C.; Cannone, J. J. *Curr. Opin. Struct. Biol.* **2002**, *12*, 301-310.
99. Trieber, D. K.; Williamson, J. R. *Curr. Opin. Struct. Biol.* **2001**, *11*, 309-314.
100. Sosnick, T. R.; Pan, T. *Curr. Opin. Struct. Biol.* **2003**, *13*, 309-316.
101. DeRose, V. J. *Curr. Opin. Struct. Biol.* **2003**, *13*, 317-324.
102. Fedor, M. J. *Curr. Opin. Struct. Biol.* **2002**, *12*, 289-295.
103. Onoa, B.; Dumont, S.; Liphardt, J.; Smith, S. B.; Tinoco, I., Jr. *Science* **2003**, *299*, 1892-1895.
104. Woodson, S. *Cell Mol. Life Sci.* **2000**, *57*, 796-808.
105. Schroeder, R.; Grossberger, R.; Pichler, A.; Waldsich, C. *Curr. Opin. Struct. Biol.* **2002**, *12*, 2960300.
106. Chiu, T. K.; Kaczor-Grzeskowiak, M.; Dickerson, R. E. *J. Mol. Biol.* **1999**, *292*, 589-608.
107. Woods, K. K.; McFail-Isom, L.; Sines, C. C.; Howerton, S. B.; Stephens, R. K.; Williams, L. D. *J. Am. Chem. Soc.* **2000**, *122*, 1546-1547.
108. Subirana, J. A.; Soler-López, M. *Annu. Rev.Biophys. Biomol. Struct.* **2003**, *32*, 27-45.
109. Sines, C. C.; McFail-Isom, L.; Howerton, S. B.; VanDerveer, D.; Williams, L. D. *J. Am. Chem. Soc.* **2000**, *122*, 11048-11056.
110. Hud, N. V.; Plavec, J. *Biopolymers* **2003**, *69*, 144-159.
111. Chiu, T. K.; Dickerson, R. E. *J. Mol. Biol.* **2000**, *301*, 915-945.
112. Hud, N. V.; Polak, M. *Curr. Opin. Struct. Biol.* **2001**, *11*, 293-301.
113. Ahmad, R.; Arakawa, H.; Tajmir-Riahi, H. A. *Biophys. J.* **2003**, *84*, 2460-2466.

114. Braunlin, W. H.; Drakenberg, T.; Nordenskiöld, L. *J. Biomol. Struct. Dyn.* **1992**, *10*, 333-343.
115. Rouzina, I.; Bloomfield, V. A. *Biophys. J.* **1998**, *74*, 3152-3164.
116. Kielkopf, C. L.; Ding, S.; Kuhn, P.; Rees, D. C. *J. Mol. Biol.* **2000**, *296*, 787-801.
117. Egli, M. *Chem. Biol.* **2002**, *9*, 277-286.
118. Ivanov, V. I.; Krylov, D. Y.; Minyat, E. E.; Minchenkova, L. E. *J. Biomol. Struct. Dyn.* **1983**, *1*, 453-460.
119. Ivanov, V. I.; Minchenkova, L. E.; Minyat, E. E.; Schyolkina, A. K. *Cold Spring Harbor Symp. Quant. Biol.* **1983**, *47*, 243-250.
120. Nishimura, Y.; Torigoe, C.; Tsuboi, M. *Biopolymers* **1985**, *24*, 1841-1844.
121. Benevides, J. M.; Wang, A. H.-J.; Rich, A.; Kyogoku, Y.; van der Marel, G. A.; van Boom, J. H.; Thomas, G. J., Jr. *Biochemistry* **1986**, *25*, 41-50.
122. Wolk, S.; Thurmes, W. N.; Ross, W. S.; Hardin, C. C.; Tinoco, I., Jr. *Biochemistry* **1989**, *28*, 2452-2459.
123. Howerton, S. B.; Sines, C. C.; VanDerveer, D.; Williams, L. D. *Biochemistry* **2001**, *40*, 10023-10031.
124. Tereshko, V.; Minasov, G.; Egli, M. *J. Am. Chem. Soc.* **1999**, *121*, 3590-3595.
125. Johansson, E.; Parkinson, G.; Neidle, S. *J. Mol. Biol.* **2000**, *300*, 551-561.
126. Buckin, V. A.; Kankiya, B. I.; Rentzeperis, D.; Marky, L. A. *J. Am. Chem. Soc.* **1994**, *116*, 9423-9429.
127. Auffinger, P.; Westhof, E. *J. Mol. Biol.* **2001**, *305*, 1057-1072.
128. Soler-López, M.; Malinina, L; Subirana, J. A. *J. Biol. Chem.* **2000**, *275*, 23034-23-44.
129. Egli, M.; Tereshko, V.; Teplova, M.; Minasov, G.; Joachimiak, A.; Sanishvili, R.; Weeks, C. M.; Miller, R.; Maier, M. A.; An, H.; Cook, P. D.; Manoharan, M. *Biopolymers* **1998**, *48*, 234-252.
130. Beveridge, D. L.; McConnell, K. J. *Curr. Opin. Struct. Biol.* **2000**, *10*, 182-196.
131. Feig, M.; Pettitt, M. *Biophys. J.* **1999**, *77*, 1769-1781.
132. Manning, G. S. *Q. Rev. Biophys.* **1978**, *11*, 179-246.

Chapter 2

Molecular Dynamics of DNA and Protein–DNA Complexes: Progress on Sequence Effects, Conformational Stability, Axis Curvature, and Structural Bioinformatics

D. L. Beveridge, Surjit B. Dixit, K. Suzie Byun, Gabriela Barreiro, Kelly M. Thayer, and S. Ponomarev

Chemistry Department and Molecular Biophysics Program, Wesleyan University, Middletown, CT 06459

Introduction

Molecular dynamics (MD) computer simulations on DNA and protein-DNA complexes are now well into second generation [1], with current calculations including solvent at ionic strengths relevant to in vitro experiments and in vivo phenomena newly feasible on the nanosecond (ns) time frame. Recent MD results are much improved over earlier reports [2-4], and now provide plausible description of all-atom structures as a function of time for DNA oligonucleotides free in solution, and complexed with proteins [1, 5, 6]. A number of problems of considerable interest in the structural biology of DNA, such as the nature of the hydration and ion atmosphere, sequence effects on structure, axis curvature and flexibility, the conformational landscape of the DNA double helix, structural adaptations of DNA on protein binding, and aspects of the relationship between dynamical structure and functional energetics are uniquely accessible to study at improved levels of rigor. Considerable advances have been made on methodological issues, particularly with respect to force fields and run lengths, although certain aspects as described herein remain to be fully secured. In this article, we provide some historical perspectives on the subject, review recent results, and provide our current opinion of the present state of play, problems and prospects. The accompanying figures, without further apology, are taken from our own recent research papers. A series of recent review articles treat in more detail the topics of molecular dynamics simulations of DNA [1, 2, 7], nucleic acid force fields [8], DNA hydration [2], the ion atmosphere of DNA [9], DNA bending [7], and protein DNA interactions [4]. Notable independent reviews of MD on DNA are due to Cheatham and Kollman [10] Guidice and Lavery [11], Norberg and Nilsson [12] and Orozco et al. [13].

14

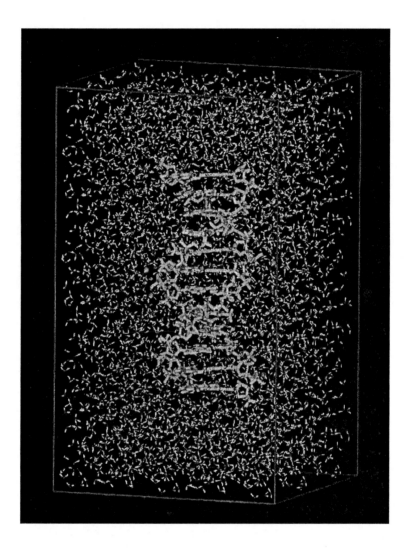

Figure 1. a) System configuration for an MD simulation on the Eco RI dodecamer duplex, d(CGCGAATTCGCG), in a rectangular cell and including counterions and water molecules. *(Reproduced with permission from reference 18. Copyright 1977 Biophysical Society.)*

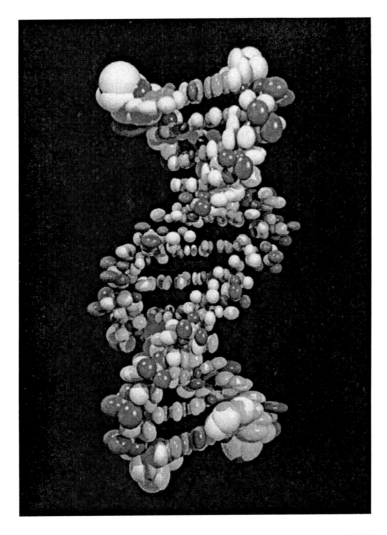

Figure 1. b) calculated dynamical structure of the Eco RI dodecamer after 5 ns of MD, presented in terms of thermal ellipsoids centered on each atom [18]. *(Reproduced with permission from reference 18. Copyright 1977 Biophysical Society.)*

Molecular Dynamics: Methodology

MD simulation procedures for DNA oligonucleotides (**Figure 1**) have stabilized somewhat in the course of the last few years. The problem of long range interactions is seemingly under control with the advent of the particle mesh Ewald (PME) method [14] for boundary conditions, despite some lingering concerns about long range correlations [15]. Current perspectives on MD force fields for nucleic acids are provided in recent reviews by Cheatham and Young [5] and by Mackerell [6] . The AMBER nucleic acids force field based on the second generation parameterization of Cornell et al.[16], known as parm.94, was used in the first well-behaved MDs on the DNA double helix with explicit solvent [17-19] and is our current method of choice. Agreement between calculated and observed results has been quite good overall, with a known apparent deficiency being a slight tendency towards undertwisting of B-form DNA as compared with standard values of 36°. We have studied base pair twist as a function of sequence and have reproduced the undertwisting but find the trends in close accord with the benchmark experimental results [20]. A modification of AMBER parm.94 known as parm.99 has been proposed [21], but the possible improvements are somewhat qualified. As refined by Mackerrel and coworkers, CHARMM [6, 22-25] provides a viable force field alternative for MD on nucleic acids, with the latest studies showing likewise reasonable accord with experiment. It is our opinion that MD on DNA, with attention paid to the minor discrepancies relevant to question being pursued, can be approached by either of these options. Further explicit comparative studies [26, 27] compare the two force field options. Feig and Pettit [26] with 10 ns MD trajectories note some salient issues and the need for longer simulations. The study of Reddy et al. [27] is based on relatively short, 1 ns trajectories. A comparison of force fields based on longer run lengths has been carried out by T.E. Cheatham (private communication).

The methodological questions typically asked of MD are a) when is a simulation "converged"?, b) what length of trajectory is "enough"?, c) how sensitive are the results to the choice of initial configuration, and d) how does one extract from MD the "structure"? Questions a) and b) are in fact moving targets with no known definitive answer, but important to consider nonetheless. Convergence in a simulation can never be unequivocally proved since in any inductive science one cannot guarantee past behavior of a system is predictive of the future – one may always encounter a new dynamical regime. With present supercomputers, one deals with this by running simulations for as long as one can afford (our longest to date is 60 ns), and checking on the stability of diverse indices of dynamical structure as a function of time. Each property or structural index exhibits a time evolution in MD. The rule of thumb is to sample for 10 times the relaxation time of all indices of interest in an application [28]. A particular convergence issue arises in systems in which the dynamical structure is comprised of a Boltzmann weighted average over a set of thermally accessible substates. An analysis technique for this involves two-dimensional root mean square distance (2D-RMSD) plots on positions, conformational or helicoidal parameters resolved into contributions from substructures. As MD on DNA samples further decades of time substates are likely to become an increasingly important issue to modelers [29-32].

DNA Structure and Dynamics.

   The discovery of the structure of DNA in 1953 and elucidation of the modus operandi of a self-replicating, informational macromolecule are the foundations of modern molecular biology [33]. The structural biology of DNA has been described extensively in various texts [34-36] and a recent multi-author volume [37]. The definitions of the essential structural parameters are reviewed there by Lavery and Zakrzewska [38]. Fiber diffraction studies have characterized the basic conformational families of right-handed DNA duplexes as A-form and B-form [39]. The B form of DNA is stable at high relative humidity in fibers and relatively high water activity in solution [34]. The B to A interconversion in calf thymus DNA in solution can be induced by simply altering water activity, indicative of the significant role that environmental effects play in DNA conformational stability. Sequences rich in CG base pairs have a propensity for B to A interconversions, whereas AT rich sequences are B-philic, an example of sequence effects on conformational stability which must be accounted for quantitatively by any successful general model [40]. Structural deformations of the double helix are implicated in DNA complexes with proteins and to some extent with smaller ligands [41, 42]. DNA in these instances remains a right-handed helix, and thus the entire range of structures located on the conformational landscape of right-handed DNA duplexes are of interest, as well as the left-handed Z-form. While a number of intermediate right-handed structures (A, B, B' C, D,...) have been proposed based on crystal structure evidence [34], just which are distinct in solution and which are not remains an open question. The B'-form of DNA [43] characteristic of sequences of adenines (A-tracts) is of particular significance with respect to concerted axis bending. Helix melting and compaction are more extreme forms of conformational transitions which are on the verge of accessibility by MD models. The stability of the double helix includes contributions from van der Waals attractions (dispersion forces) and repulsions (steric effects), electrostatic interactions between bases and base pairs (hydrogen bonding and stacking effects), and solvent effects which also include hydrogen bonding, electrostatic polarization of solvent and hydrophobic bonding as well as interactions involving mobile counterions and coions. A fully unequivocal treatment of DNA sequence effects on structure based on these contributions has not yet been achieved, but considerable progress has been reported (see below).

   The first "glimpse" of the right-handed DNA double helix at or near molecular resolution was obtained from the crystal structures of the dinucleotides which exhibit Watson-Crick Base pairing [44] and the oligonucleotide dodecamer duplex d(CGCGAATTCGCG), which crystallized in the B-form [45, 46]. This sequence contains the recognition site for the restriction enzyme Eco RI endonuclease and was a component of the first protein-DNA crystal structure [47] and an early drug-DNA complex [48]. The helicoidal irregularities observed in the structure raised the issue of sequence effects and the idea that the conformation of the molecule might be part of the recognition code for sequence specific interactions of DNA with ligands [49]. A series of ordered solvent positions was identified in the minor groove region of the AATT tract and termed the 'spine of hydration' [50]. This entity was suggested to be specific for AT regions and a determining factor in the preferential stability of the B form of DNA under conditions of high humidity [51]. The crystal structures of DNA oligonucleotides are a valuable source of structural data [52, 53] and are used extensively to catalog and develop hypotheses about sequence effects on structure (see below) which may judiciously be applied to the solution state as well. However, the extent to which details of these structures are influenced by crystal packing effects remains a matter of debate [54] - in some cases two conformers of the same molecule are seen in a crystal [55, 56], and there was a concern that crystalline environment alters axis bending even for short sequences [57]. The latest results from MD on DNA have been tested against experimental data on only a limited number of crystals [1, 58, 59]. In the crystalline state, the HPV E2 DNA oligonucleotide sequence, d(ACCGAATTCGGT), exhibits three different structural forms [56]. Studies of the structure of E2 DNA in solution based a series of MD simulations utilizing both the canonical and various crystallographic structures as initial points of departure [60]. In 4 nanosecond trajectories, all MDs converged on a single dynamical structure of d(ACCGAATTCGGT) in solution. The predicted structure is in close accord with two of the three crystal structures, and indicate that a significant kink in the double helix at the central ApT step in the other crystal form may be a packing effect. The calculated curvature in the sequence was found to originate primarily from YpR steps in the regions flanking the central AATT tract.

NMR spectroscopy provides NOE intensity data [61] and now residual dipolar couplings (RDC) [62, 63] as a basis for determining DNA structures in solution. Details of the advances in DNA structure determination using RDC are described in two recent papers [64, 65]. However, the NMR structures are reported as a bundle of structures consistent with the data but not a true Boltzmann ensemble, a matter that raises further theoretical issues. The latest results from MD on DNA for d(CGCGAATTCGCG) tested against NMR solution structures [1, 66], indicate that MD provides a reasonably accurate description and plausible model of the dynamical structure of DNA oligonucleotides in solution. In a key NMR structure determination of d(GGCAAAAAACGG), in which the results differs from that found in the crystal, MD beginning at NMR structure, canonical B-form or the crystal structure supports the NMR solution structure, and provides further indication of

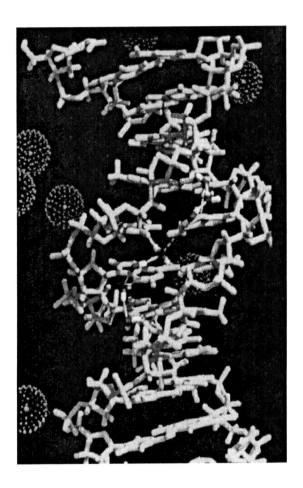

Figure 2. Snapshots from an MD trajectory on the Eco RI dodecamer, looking into the minor groove. A) an example from MD of an intact spine of hydration. B) an example from MD of a Na$^+$ ion showing transient occupancy in the ApT pocket of the minor groove. [19]. *(Reproduced from reference 19. Copyright 1977 American Chemical Society.)*

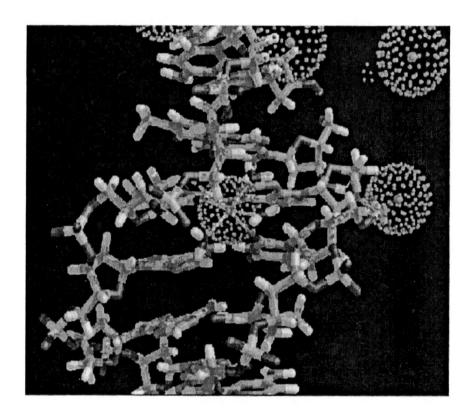

Figure 2. *Continued.*

the extent that the direction of curvature in crystal structure may be subject to packing effects [67].

DNA Solvation

Hydration. Water vapor sorption isotherms for a number of nucleic acids were reported in the early 1960's by Falk et al. [68-70] and used to describe the apparent stoichiometric equilibrium between water "bound" to DNA and that otherwise situated in the system. These results and changes in the infrared absorption spectra on dehydration led to a proposed hydration number of 25-27 waters/ nucleotide for DNA samples at high water content [71]. The strongest water binding was identified with the 5-6 waters/ nucleotide assigned to phosphate groups, followed by hydrogen bonding of water to the donor and acceptor hydrophilic sites of the nucleotide bases presented to the major groove and minor groove of the double helix. Sugars were expected to be the least hydrated, but may figure into hydrophobic hydration in a significant way. The recent literature [72] contains many references to special water structures around DNA including bridges, tridents, cones, pentagons, spines, trans-groove filaments, and clathrates in addition to variations on the tetrahedral coordination of water exemplified by ice Ic. Some of these are invoked in hypotheses about the conformational stability of DNA, e.g. the role of the minor groove spine of hydration (**Figure 2a**) in the preferential stabilization of B-form DNA at high water activity [51], a water bridge in the "economics of phosphate hydration" argument invoked to explain the transition from B to A as water activity is lowered [73], and "tridents" in DNA bending [74]. A problem in the field is that the direct causal link with the free energy of stabilization contributed by these hydration substructures has never been securely made, although some proposals are plausible. Another problem is that hydration together with ion solvation of the double helix is robust, whereas the crystallographically ordered waters comprise only a small fraction of the total solvation. Disordered waters contribute to the thermodynamic stability of the system, and must be considered in a proper theoretical model. The direct connection between crystallographically observed or theoretically postulated water structure and thermodynamic stability remains a missing link in the DNA hydration field. A recent review of DNA hydration by Pettitt [75] draws the useful distinction between "solvation site" and "distribution function" approaches. Crystal structure articles reporting ordered water positions have been analyzed in depth for nucleic acids by Schneider and Berman [76, 77], who delineate characteristic transferable features of the hydrophilic hydration sites of nucleotide bases and the "cones of hydration" around anionic phosphates. MD calculated distribution functions presented via computer graphics as hydration densities provide a more complete (albeit theoretical) view of the hydration, and can be decomposed by the proximity method [78, 79] to arrive at useful interpretations. This method was extended in several recent papers by Pettitt and coworkers

[80]. A calculated solvent structure of the primary solvation shell of DNA sequences was compared with the location of ordered solvent positions in the corresponding crystal structure by Feig and Pettitt [81] with considerable success. Pettit and coworkers [75] have just definitively reviewed the field.

Counterions and Coions. DNA occurs at physiological pH as a polyanion, the acidic phosphate groups being fully ionized. Electroneutrality demands at least an equivalent number of counterions to be present in the system which, being mobile, assume a statistical distribution in the space proximal to the DNA. Additional ions from any excess salt in the system together with the original counterions form the total ion atmosphere, which along with water constitutes the DNA solvation. DNA structure is observed to be sensitive to the composition and concentration of the ion atmosphere as well as the water activity, most dramatically illustrated by the change in helix sense involved in the transition from right-handed B-DNA to left-handed Z-DNA at high salt in GC rich sequences [82]. Diverse less global changes are observed for various (and important) sequences. A phenomenological model for the ion atmosphere of DNA is "counterion condensation" [83]: no matter how dilute the solution, a number of the counterions remain in close proximity to the DNA, compensating a large percentage of the phosphate charges and said to be "condensed." Elaboration of a more general PB model of ion atmosphere has been advanced in the thermodynamic binding and NMR studies of Record and coworkers [84]. Limitations of the PB theory due to neglect of finite size of the mobile ions and spatial correlation have been a matter of some concern, and have been characterized by comparisons with Monte Carlo calculations [85-87] and with hypernetted chain theory [88]. General accord has been established, although there is indication that the PB approach may underestimate the concentration close to the DNA [86]. Both PB and MC calculations have been used to investigate Manning's hypothesis about the insensitivity of charge compensation to concentration, and found in fact slight but potentially significant variation [87]. The behavior of counterions around DNA and nature of the counterion atmosphere at the molecular level remains a developing subject. Early NMR experiments were cited in support of the idea that all small cations are in a state of complete hydration with at least one dimension of translational mobility, i.e. delocalized and rather loosely associated with the DNA. The Manning radius (of the counterion condensate) is ~12 Å [83] and quite possibly larger [88]. On the other hand, DNA duplex rotation angle has been found to vary systematically with cation type [88], which may require an explanation involving some degree of site binding. Multivalent cations are more disposed to site binding than monovalent ions [88]. The NMR literature is quite extensive on cation resonance in DNA systems [89, 90], but the data has proved to be difficult to interpret unequivocally in terms of structure.

Release of condensed counterions is an important thermodynamic component of ligand and protein binding to DNA [91] and has motivated further studies of structure of ions around DNA. Analysis of the distribution of mobile counterions from MD have produced an independent description the DNA ion atmosphere (**Figure 3**) [19, 92]. The results exhibit features in reasonable accord with the core concept of "counterion condensation" of Manning theory [83], and provided an independent account of the Manning fraction for monovalent counterions. A surprising (ca. 1997) result came from the analysis of solvent around the EcoRI dodecamer: a Na^+ ion found to assume some fractional occupancy in the minor groove at the ApT step, indicating counterions may intrude into the venerable spine of hydration [19] (**Figure 2b**). This "ApT pocket" was noted previously to be of uniquely low negative electrostatic potential relative to other positions of the groove [93] and supported by the location of a Na^+ ion in the crystal structure of the rApU mini-duplex [94]. Support for the fractional occupation of ions in the minor groove of B DNA was subsequently obtained by NMR [95, 96] and crystallography, particularly in the cases of K^+, Rb^+, and Cs^+ salts [97, 98]. Other MD studies focusing on the ionic atmosphere of DNA have been carried out by Lyubartsev et al. [99] who compare the behavior of Cs^+, Na^+, and Li^+ ions and find Li^+ ions to be primarily coordinated to the phosphates but not Cs^+ and Na^+. The MD results were used to help interpret NMR data [100] showing Li^+ to have a much slower diffusion rate compared to Cs^+.

The idea that sequence dependent concentrations of ions can influence structure [19, 101, 102] has received considerable subsequent attention [98, 103, 104]. Young et al. [19] have provided a schematic of possible ion binding sites in the major and minor groove of DNA (**Figure 4**) which indicate possible major groove as well as minor groove electrostatic pockets. Williams and coworkers [103], based on the idea of structural deformations arising from the modulation of phosphate repulsions [105], have argued the case for "cations in charge" of DNA structure. This idea has been recently argued [106], based on the crystal structure of a cross-linked dodecamer in environmental conditions without appreciable monovalent cations. The structure was not significantly altered, indicating that either ions were not present in the spine or, if present, were not causing structural perturbations or "electrostatic collapse." Williams and coworkers [97] have subsequently examined the issue of groove width and ion occupancy in, as Egli and coworkers [98] call it, the "hydrat-ion" spine of the minor groove, and the evidence for a significant presence of ions at favorable sites in the minor and also major is mounting [104]. Stellwagen et al. [107] present experimental data in support of this based on effective charge of A-tract oligomers based on capillary electrophoresis. So far MD results are supportive of the idea of ions in the grooves, but the extent of the effect and implications are as yet unsettled. Based on a restrictive definition of groove width, we found the fractional occupancy of ions in the minor groove to favor

24

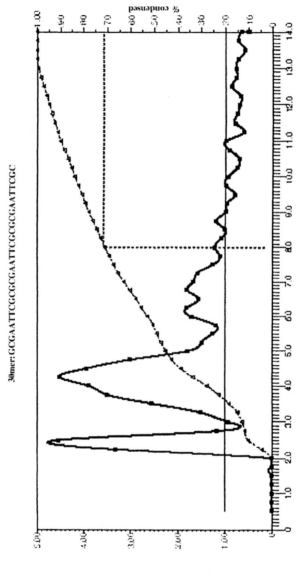

Figure 3: Proximity-based radial distribution function of Na$^+$ ion density as function of the distance from the atoms of the central 10 base pairs in the DNA duplex d(CGCGAATTCGCGCGAATTCGCGAATTCGCG) are shown. The left axis is the relative probability. Dashed straight lines indicate running Na$^+$ coordination, with an inflection point in the region of the Manning radius, ~ 8.0 – 10.0 Å. Triangles indicate the running number of Na$^+$ atoms relative to the total charge on the DNA (right axis) [5]. This analysis shows the extent to which the results of MD simulation are consistent with the phenomenological counterion condensation model [83] of the ion atmosphere of DNA.

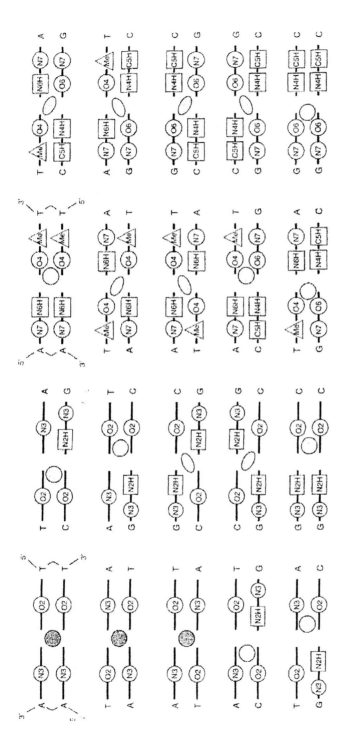

Figure 4: a) Schematic view of DNA base pair steps in the minor groove indicating possible locations of electrostatic pockets, indicated by shaded regions [19]. The shading is proportional to the estimated efficacy of the site for binding cations. Electronegative atoms On the base pair edges are indicated by circles, electropositive hydrogen atoms by squares, and hydrophobic methyl groups by triangles; b) Schematic view of possible electrostatic pockets of DNA base pair steps in the major groove.

ApA and ApT steps of the minor groove and TpA in the major groove, with major groove occupancies typically larger as a consequence of the increased amount of configuration space as well as electronegative pockets. Williams and coworkers [108] based on a definition which extends to the solvation of phosphates, argue for a correlation between groove width and ion occupancy. However, conclusions about influence of mobile ions on DNA groove widths at the level of MD in the latest report by Williams (10 ns) and this group (now 60 ns) are not in full accord. Our results are based on simulations on a palindromic DNA oligonucleotides, in which the symmetry mandates equivalent ion distributions around each strand of the duplex. Previous results were based on 5 ns trajectories [5]. Our most recent results indicate that the equivalence of ion distributions of the two palindromic strands of DNA is only at the level of r = .64 at 14 ns (**Figure 5**), but some stability in the site positions can be demonstrated. Pushing the trajectory to 60 ns results in a correlation of ion occupancies on the two strands to be ~.94 which extrapolates to ~.98 at 100 ns [109]. The structural parameters of the DNA converge much faster (3-5 ns). A comprehensive treatment of this issue and the implications thereof with respect to recommended simulation protocol is in progress here. The solvent part of the problem is expected to be even slower to converge when divalents are involved, and these will be important for nucleic acid problems such as DNA compaction [110]. This has implications in conductivity studies of DNA, since even fractional occupation of ions in the grooves of DNA could influence electron transfer and hole migration [111]. All these studies presume an accurate set of ion potentials. The current Aqvist parameters [112] derived from free energy simulations are widely used, and current issues have been delineated by Roux [113]. Polarization effects can be expected to be important, especially for divalents. The current state of polarizable force fields has been reviewed by Halgren [114]; a production version of a polarizable force field for nucleic acids is not yet available.

Base Pair Sequence Effects on DNA Structure

The sequence dependent structural irregularities in the DNA double helix are a potential Rosetta Stone for understanding conformational stability, the affinity and specificity of DNA ligand interactions, but just as elusive. Research efforts to date aimed at understanding the sequence effects are focused primarily on a basis set of all possible dinucleotide steps, which for A, T, C and G bases present 16 possibilities of which 10 are unique. These may be grouped into the classes YR, YY (=RR) and RY where Y stands for pyrimidine bases and R for purines. With respect to molecular geometry, the problem is reduced to the subset of helicoidal parameters which show significant variability over the data base. Recent works have focused on the propeller deformation of WC base pairs and the base pair step parameters roll, twist and slide [115]. In a most comprehensive mining of the crystallographic data base to date with respect to

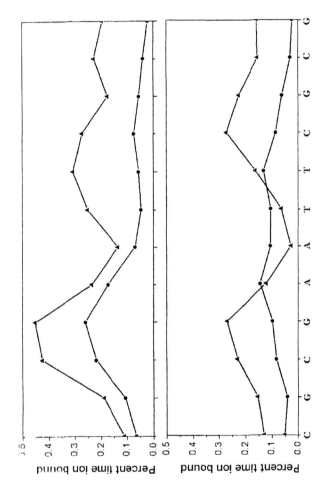

Figure 5. NA$^+$ Ion occupancies per DNA base pair. Minor (triangles) and major (circles) groove occupancies as a function of sequence calculated from 14 ns (top) and 60 ns (bottom) of the simulation on the Eco RI dodecamer. Due to the palindromic symmetry of the sequence, fully converged ion distribution results should exhibit symmetry with respect to the midpoint of the dodecamer, the ApT step.

classes, Suzuki and coworkers [116] show that a) A-form oligonucleotides show relatively little differentiation with respect to sequence and b) B- form oligonucleotides show clear evidence of sequence effect, with values for YR, YY and RY steps forming essentially non-overlapping clusters in roll/slide/twist space. Extending this resolution further into A, G, C and T reveals additional interesting patterns as elaborated by El Hassan and Calladine [117], who classify base pair steps from oligonucleotide crystal structures into the categories rigid (AA/TT, AT, GA/TC), loose (GG/CC, GC, CG, CA/TG, TA, AC/GT), bistable (homogeneous G,C steps), and flexible (CA/TG, TA). Subsequent analyses of the crystal structures are due to Liu et al [118], and Olson and Zhurkin [119].

There is a developing literature of post hoc explanations of these patterns in terms of chemical forces. In a series of articles and summarized a their recent text, Calladine and Drew [36] advanced an argument that originates in the observation that AT pairs exhibit significantly higher propeller than CG pairs. Another observation from crystallography is that AT sequences and particularly runs of adenine nucleotides (A tracts) are relatively straight (little deviation from B) and rigid (limited dispersion in geometry). Some controversy exists over the origin of this effect, i.e. whether it is due to an additional "bifurcated" H-bond between A-N6 and T-O4 across a step or occurs as a consequence of steric clashes due to the large propeller; MD and experiment seem to disagree on this issue [120-123]. Suzuki et al [116] position their analysis on the base pair step level, and show YR and RY steps to present two extremes, all positive roll/ slide/ twist, and all negative. Explanations for the trends and correlations are advanced based on a strictly steric, trapezoidal slab model of AT and CG base pairs. GC stacks show evidence of being bistable with values characteristic of both A and B forms. Hunter [124] has introduced some quantitation into the problem based on simple QM vs. steric calculations and observes the electrostatic effects are more likely in stacks of GC pairs due to a non-negligible dipole moment as compared with AT pairs. Friedman and Honig [125] and Elcock and McCammon [126] introduced solvent effects quantitatively into the sequence effect problem.

MD calculated sequence effects for the 10 unique base pair steps in DNA oligonucleotides are being used as a basis for a more detailed comparison of theoretical and observed values, and provide a higher resolution characterization of force fields (**Figure 6**). Results from B-DNA crystal structures [52] have been collected as a function of the helicoidal parameters roll, twist and slide [116]. The YR, RR (=YY), and RY steps over all B-form DNA crystal structures are found to fall into fairly distinct clusters. The corresponding results from MD simulations on DNA sequences including water and counterions using AMBER and the parm.94 force field have been compared [1, 122]. The results general follow the trends identified by Suzuki et al., with ApA steps essentially straight, YpR steps strongly favoring deformation toward the major groove, and RpY steps showing a preference for deformations towards to minor groove, but less emphatically. The results show a systematic displacement in slide in the calculated values compared with the observed, indicating a possible systematic force field discrepancy. The implications of this are not fully clarified, but slide makes a critical differentiation in A-form vs. B-form DNA in helicoidal parameters derived with respect to a local, base pair centered, frame of reference [127, 128]. An emerging issue with respect to dynamical structure is the population of substates of YpR sequence steps, in which open and closed state of a hinge-like structure may play a key role in DNA curvature and premelting phenomena (see below). The issue of how well sequence effects are accounted for in our best current all-atom MD model is a vital issue in MD on DNA. The basic idea is that one must know the dynamical structures of all ten unique base pair steps in all possible sequence contexts. This requires structures of 136 unique tetrameric steps [129-131]. However, the crystallographic data base includes examples of only a fraction of these, and NMR RDC structures are available only for a small number of cases. Thus the basic question about sequence effects on DNA structure as well as the underlying question of which structural factors are responsible remains to be fully clarified. Here MD has a vantage point in that obtaining predictions about the dynamical structures of tetramer steps is currently feasible, albeit a large computational undertaking. A consortium of investigators has taken up this problem, and initial results at the level of 15 ns trajectories on all tetranucleotide steps are complete and being prepared for publication [132].

DNA Conformational Stability and Structural Transitions

Early studies of DNA using fiber diffraction [133] resulted in the recognition of hydration dependent forms and the now familiar assignment of the conformational labels A and B to fiber forms stable at low and high relative humidities, respectively. Subsequently, Wolf and Hanlon [134] and Ivanov et al [135] showed that similar conformational behavior occurs in solution as a function of water activity. The A to B transition in DNA is observed to be sensitive to the base sequence, and C/G base pairs are generally more A-philic than A/T base pairs [43, 136]. Several groups have shown the addition of salt as well as organic cosolvents cause B to A transitions [137, 138]. Conventional wisdom holds the issue is basically a matter of water activity rather than the specific nature of the ions or cosolvent, but further studies are in order on this point.

Structural deformations of B-form DNA observed in ligand binding occur mainly on the conformational landscape of right handed DNA structures [119]. Thus detailed studies of B vs. A conformational stability and B to A conformational transitions are an essential target of MD models. Salt effects are an additional complicating factor, and multivalent cations are known to be important in DNA compaction [110]. Systems with solvent water, counterions and coions at experimental and even physiological ionic strengths can now be

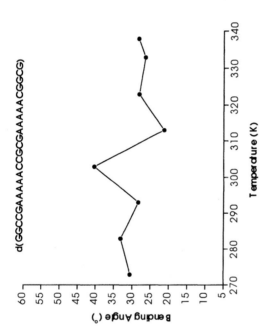

Figure 6a – Calorimetric and spectroscopic experimental results indicated that DNA sequences with A-tracts phased by a full turn of a B-form helix present an unusual structure at low temperatures, the B' form, and show a B' to B form premelting transition, ~ 300-312 K [145]. This plot of bending angle (°) versus temperature for the phased A-tract sequence d(GGCCGAAAAACCGCGAAAAACGGCG) . The mean values are the bending angle average values from a 3ns MD simulation at eight different temperatures. The discontinuity at ca. 313 K corresponds to a B' to B DNA conformational transition.

31

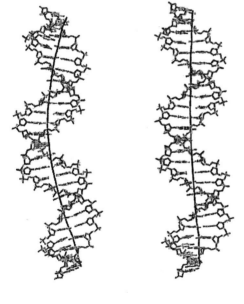

Figure 6b – MD average structures and calculated bending angles for d(GGCCGAAAAACCGCGAAAAACGGCG) based on a 3ns trajectories at 303 and 313 K. Molecular dynamics simulations, with explicit water and counter ions, indicate this sequence to be more bent at temperatures below the premelting transition. Below the premelting transition, the structure is divided into B' and B forms, which establish a phasing frame and amplifies the concerted axis bending. Above this temperature the structures are relatively straight. With no B'/B differentiation, there is no structural helix phasing, and the bending enhancement disappears.

treated in MD and make interesting problems in the area of conformational stability and transitions accessible to more rigorous theoretical study. Cheatham and Kollman [139] reported MD on d(CCAACGTTGG) in which the expected A/B conformational preferences were observed in water, including an A to B transition. However, at low water activity, a simulation beginning in the B form did not make the expected conformational transition to the A in a trajectory of 500 ps. Free MD on the sodium salt of the Eco RI dodecamer in water and in a mixed solvent comprised of 85% (v/v) ethanol water were reported by Sprous et al [140]. This sequence is observed to assume a B-form structure in the solid state and in aqueous solution and is expected to assume an A-form structure in the mixed solvent environment. Three additional simulations are reported: one simulation starts from the A-form in water, the second starts from the A-form in 85% (v/v) ethanol:water and the last starts from the B forms in 85% (v/v) ethanol:water. The MD on A-form structure in water undergoes an A to B-DNA transition and stabilizes in the B-form. The corresponding 2.0 ns MD in ethanol:water remains an A-form structure, as expected. However, the B-form structure in the 85% (v/v) ethanol:water remains B-form even after 2.0 ns of MD. In a follow-up studies [40, 141], minor modifications of a single dihedral force constant in the sugars enabled the MD structure to convert to A, but a decline in agreement on issues on the preferential stability of B form in water were noted. The role of sugar repuckering in the A to B transition was investigated by Solvia et al [142], demonstrating that a driven change of one dihedral angle in the sugar ring leads to a fast, complete change in the DNA conformation. With three sets of MD simulations of the A to B transition reported [139, 140, 143], a problem is that in all cases the model sequence, while featuring reduced water activity, does not correspond 1:1 to experimental studies. Future studies in this problem sector need to be better defined and addressed to specific sequences and systems which are well characterized experimentally. The above set of MD simulations formed the basis for a subsequent free energy analysis of the conformational preferences of A and B forms of DNA in solution as a function of water activity [40, 144] and possible explanations for the origins of the environmental sensitivity of DNA conformation.

Structural deformations of B-form DNA observed in ligand binding occur mainly on the conformational landscape of right handed DNA structures. Thus detailed studies of B vs. A conformational stability and B to A conformational transitions as well as the more nuanced B' to B premelting transition are an essential calibration point for MD models. At least three sets of MD simulations of A to B transition have been reported [139, 140, 143]. To date, MD models of A to B transition show the process to be spontaneous on the ns time scale in water. However, the corresponding B to A transition in ethanol/water was not spontaneous, at leased based on AMBER parm.94. This has been interpreted as a force field problem, and indeed a minor reparameterization of parm.94 to parm.99 was engineered to make B to A more favorable but this particular modification created incipient rigidity elsewhere in the MD model [21]. We have been concerned that the behavior expected from the simulation was based on generic expectations about rather than those specific to a given oligonucleotide sequence. In one case in particular (our own), the B to A transition in mixed ethanol water solvent was studied in the absence of data explicitly showing this sequence adopts the A form under these conditions, although that would be generally expected[34]. In collaboration with Prof. Ishita Mukerji and coworkers, we have initiated a joint project to establish A/B stability as a function of water activity and salt for a set of specific oligonucleotide sequences to serve as a well defined point of comparison for corresponding MD studies as well a basis for more detailed considerations of sequence effects. The experiments involve both CD and Raman spectroscopy to monitor A/B concentrations. Mukerji et al. confirmed that d(CGCGAATTCGCG) does indeed adopt the A form in mixed ethanol water solvent, at least putting that issue to rest. At this point we propose to generate a full A/B stability diagram as a function of water activity and ionic strength for specific sequences, d(CGCGAATTCGCG) to start with, perform corresponding MD simulations on the exact same sequence and conditions of solvent and ionic strength, and use this to obtain a more rigorous basis of assessment of the performance of corresponding MD simulation.

It is well known from both crystal structures and footprinting experiments that A-tracts in DNA oligonucleotides exhibit a variation of the canonical B-form structure with narrow minor groove, high propeller, and slight base pair inclination known as B' DNA. Calorimetry and UVRR spectroscopy have characterized a premelting transition in A-tract DNA in which the system undergoes a transition below the melting point. We have recently made an extensive study of A-tract DNA oligonucleotides and found that the features of B'-DNA were well described by the MD model [122]. As a further sensitive test of the ability of MD to treat DNA conformational stability problems, a study of the temperature dependence of the structure of the phased A-tract sequence d(GGCCGAAAAACCGCGAAAAACGGCG), (the subject of previous calorimetry experiments [145]) , along with a control are in progress. Preliminary results indicate that the MD model actually shows a premelting transition at ~ 300°-310°, in close accord with calorimetry (**Figure 6a**). The results on a non A-tract control sequence do not show this, and thus we have a instance of MD providing a theoretical account of a extremely subtle conformational transition as a function of temperature in right handed DNA. The MD modeling provided the novel result that concerted overall bending in DNA structure decreases as a function of temperature as a consequence of a B' to B transition (**Figure 6b**) as expected but implicates as well a thermally induced shift in populations from the open to closed state of key YpR hinge steps.

DNA Axis Curvature and Flexibility.

Intrinsic DNA curvature can be determined using a number of experimental techniques, including polyacrylimide gel migration [146, 147], cyclization assays [148], electron microscopy [149] and electric birefringence [150]. New spectroscopic methods such as FRET [151-153]], LRET [154], and TEB [155] are being applied to this problem. Significant issues in the structural biology of DNA from the nature of nucleosome structure [156] to fundamental aspects of genetic recognition at the molecular level [157] are all impacted by DNA axis curvature and flexibility. Sequence dependent DNA bending results in a ten-thousandfold compaction in the nucleosome. Local deformations, in some cases extreme, are adopted by DNA in complexes with proteins involved in sequence specific regulation of transcription, translation and replication. One of earliest observed and most notable instances of sequence effects on structure occurs in nucleosomal DNA, in which two turns of DNA are supercoiled around an octamer of histone proteins [35], the subject a recent crystal structure [158]. In some cases the base pair sequence of the DNA in nucleosomes was found to have a distinct feature: tracts of adenine nucleotides (A-tracts) spaced by the number of base pairs in a full turn of the B form DNA duplex [159]. The phasing of A-tracts by a full helical turn in a DNA duplex was observed to constructively amplify bending in a concerted direction. The link between A-tracts and DNA bending has been supported by a number of experiments on DNA oligonucleotides of diverse lengths [160]. Much of the current data on DNA bending in solution observed in gels or cyclization experiments has been successfully correlated with structure via regression analysis. Bolshoy et al. [161] reported results consistent with wedge model for ApA steps. This is inconsistent however with the results obtained for base pair steps in oligonucleotide crystal structures, in which ApA steps are essentially straight in all cases (of unbound sequences) reported so far. However, crystal structures are available only on relatively short oligonucleotides and are subject to packing effects which may [55, 162] or may not [163, 164] influence the results. Liu et al. [118] concluded from a subsequent study using simulated annealing that the Bolshoy et al. fit of the bending data is not unique, and present a case that the bending data and crystal structures are not in conflict if the intrinsic dispersion of structural variables is taken into account.

The various ideas about how DNA bending relates to A-tracts can be categorized into "straight A-tract" and "bent A-tract" models. In the former category is the junction model, which holds that A-tracts and random sequence DNA are both essentially straight and that bending occurs at the junction of these two forms. Subsequent variations on the idea of straight A-tract models have been proposed, such as the bent general sequence model [165] and bent non A-tracts model [166], which postulate that bending occurs not just at junctions but more generally throughout the non-A regions of a sequence. An alternative class of models features bending within A-tracts. Here each ApA step is bent by a "wedge angle," a combination of base pair roll and tilt. In the wedge model [167] ApA steps are each bent by a small amount and random sequence DNA is presumably straight or averages to straight. Adding to the confusion is the recent reports that the rigidity of ApA steps in crystals may be sensitive to the effects of organic solvents such as MPD, a common co-crystallizing agent in the preparation of samples for X-ray diffraction studies [57]. However, the nature of this effect is disputed [168] and an explanation as to why this should preferentially affect ApA steps and not steps involving other nucleotides has not yet been successfully advanced. A key recent paper on DNA bending by interactions with multivalent cations is by Rouzina and Bloomfield [169]. They propose implication of "bending polarons" in the phenomena. This hypothesis involves self-localization of individual cations at the major groove entrance accompanied by collapse of the groove and DNA bending towards the cation and is driven by nonbonded electrostatic attraction between the compact cationic charge and the anionic phosphates from both strands of the incipiently collapsed groove. Some of the latest on this issue is the flexible ionophore model advanced by Hud et al. [104] in which A-tracts are sites of significant ion occupancy in the minor groove and G-tracts are sites of significant ion occupancy in the major groove, where the G-O2 atoms form a electronegative pocket (c.f. **Figure 4b**). Their idea is that A-tracts assume a B' form and induce bending toward the minor groove, G-tracts deform in the direction of A-DNA (the A'-form) and bend toward the major groove, and in certain sequences DNA bending is a "tug of war" between these two tendencies when both are present [170]. To our reading this may be a direct effect, or else a reason why A-tracts and G-tracts are differentiated in structure from normal B form, establish a phasing frame, and thus enhance bending which actually originates elsewhere in the sequence.

A series of recent MD papers have addressed the question of the origin of axis bending, focusing on the prototype case of ApA steps, A-tracts and A-tract induced bending [122, 171, 172]. MD studies of diverse A-tracts support the idea these structures are essentially straight in solution, in accord with results on crystal structures. MD simulation successfully describes the phenomenon of enhanced axis curvature in phased A-tracts, even under minimal salt conditions [173]. The A-tract and the non-A-tract regions together comprise the phasing element, not just the A-tract. Below the premelting temperature, the structure of A-tracts in all MD simulations to date corresponds to essentially straight, B'-form DNA. The MD results are consistent with diverse results in the literature that suggest that YpR steps are potentially flexible hinge points. The hinge involves two substates, a locally A-form (high roll, deformed towards the major groove) and locally B –form (low roll, essentially straight). Slide is not as variable in the MD as in Hunter's theoretical analysis [124], and shows up slightly negative. The MD results indicate that the open hinge substate of YpR steps, when present in a sequence and thermally populated, is the motif most likely to serve as the origin of axis curvature in DNA. Such structures are subsequently referred to as curvature elements in the helix phasing phenomena. In the frequently encountered case of A-tracts combined with a flanking sequence containing YpR steps, the relatively straight, rigid A-tracts are not an origin of bending per se but are integral to the phenomenon. The MD results suggest that A-tracts acting as positioning elements which make helix phasing as precise as possible, the result of which is maximum concerted curvature. A sequence of ordinary B- form DNA is typically more flexible than A-tracts (B'-form) and the idea is that this introduces sufficient imprecision (analogous to noise) into the helix phasing phenomena as to mitigate the amplification effect. In summary, the results of MD studies to date on the DNA, albeit based on only a few cases, account well for the essential features of DNA curvature and flexibility observed experimentally and provides a model that successfully integrates helix phasing of relatively rigid A-tracts and the YpR flexible hinge. MD provides independent support for the non A- tract model of DNA curvature [174-176] and given a full articulation by Dickerson and coworkers [177]. The junction model is a special case of YpR elements of curvature located at or near 5' junctions of A-tracts. The MD results, in accord with crystallography, do not support the ApA wedge model or a significant contribution from curvature within A-tracts as the origin of DNA curvature in sequence of phased A-tracts. Since the present article was completed, integrative research on MD simulations of DNA curvature and flexibility has been carried out and combined with a comprehensive review of MD studies of DNA curvature and flexibility [173].

Protein-DNA and Protein-RNA Complexes.

The study of protein DNA recognition was pioneered by Zubay and Doty, [178] with their insight on the complementarity of the protein α-helix and the major groove of DNA, together with the early inferences made from crystal structures by Warant and Kim [179, 180], Mirzabekov and Rich [181], Carter and Kraut [182]and the first delineation of digital readout via donor and acceptor H-bond sites by Seeman et al [183]. There are now in the hundreds of crystal structures of protein DNA complexes reported [184]. Analysis of these results has yielded a typology of binding motifs [41], the idea that protein DNA recognition involves both a digital code residing in non-covalent binding sites in the grooves of DNA (direct readout) and an analog code (indirect readout) [49] residing in the sequence dependent conformations of DNA. Useful generalities about the nature of the protein-DNA complexation are emerging from broad based analyses of the crystal structure data by Dickerson and coworkers [185, 186], Nadassy et al. [187] and Jones et al.[188] From this, a taxonomy of the functional energetics can be deduced as a basis for qualitative inferences about binding. Early molecular modeling studies on protein DNA complexation are due to Zakrzewska and Pullman [189] based on molecular mechanics and this laboratory using MD [190] which produced a first glimpse of the dynamics of structural adaptation phenomena at the molecular level, a subject that remains today an active area of research interest [191]. The functional energetics has been studied in detail on several prototype cases, particularly on the Eco RI endonuclease by Jen Jacobson [192, 193] who makes the general observations that measured affinities are ~12-15 kcal/mol for specific binding and 6-7 kcal/mol in the case of non-specific binding, and has identified the factors involved. However the relative contributions of the various factors are not directly accessible to experimental determination, which led us to a studies of protein DNA complexation based on free energy component analysis [194], aimed in part at characterizing just how much one could and could not infer from calculations of this genre. A theoretical protocol for free energy component analysis of protein DNA complexes was described first on Eco RI endonuclease as a case study and then extended to calculations ion the binding affinity of 40 protein DNA complexes based on crystal structures[194, 195]. One thing is clear: studies of complex per se are not sufficient to characterize the system. One needs a knowledge of the unbound protein and DNA as well, i.e. a full description of the initial and final states of complexation in order to understand the dynamics and thermodynamics.

With computer size and speed now able to support more computationally intensive and broad based MD modeling, protein DNA complexation presents an attractive opportunity for significant contributions to the field. The MD studies on protein DNA complexes in solution to date, exemplified by the work of Nilsson and coworkers [196-198] has been directed more at specific systems rather than general principles, which require a larger data bases of computational cases. The literature on molecular modeling studies of protein DNA complexes has recently been reviewed by Zakrzewska and Lavery [199], who concluded that beyond the binding motifs, the small number of amino acid residues involved, and general knowledge of the chemical forces, the problem is one of complex combinations of subtle facts, some indirect, which do not necessarily fit simple pairwise additive models which dominated early ideas on recognition, affinity and specificity. All atom MD studies have a unique vantage point on such problems. Protein-DNA interactions are at the crux of many important events in molecular biology, not the least of which is regulation of gene expression. From the surveys of structures solved to date, there has not yet been much success at finding a universal recognition code for protein DNA interactions at a level of simplicity and elegance of the genetic code, in fact various systems are observed to utilize the same binding motif in different ways. Matthews [200] questions whether a code exists or not, but at the same time life processes rely on high resolution specificity of interactions in order to function reliably. We are likewise intrigued with the ideas on a probabilistic recognition code by Benos et al.[201] The interpretation of results to date in terms of structural features and chemical forces, i.e. the functional energetics and molecular basis of protein DNA recognition, remains an outstanding issue of considerable consequence.

The current evidence on protein-DNA recognition speaks for a special role for structural adaptation [191], and the role of metastable states preorganized for binding [202]. The role of solvent release [203, 204] is important to clarify further at the molecular level. All atom MD studies have a unique vantage point on such problems. An initial round of MD simulations have been performed on λ repressor-operator [205-210] CAP-DNA[211] (**Figure 7**) as well as U1A- RNA [212]. Recent MD studies of the CAP-DNA complex serves as an illustrative case in point. The initial crystal structure CAP-DNA complex exhibits a dramatic instance of DNA bending, in which the DNA oligonucleotide the structure is bent by ~ 90° overall, with the origin of bending being two concerted ~ 45° kinks at YpR steps at or near points of protein DNA contacts. This system has been elaborated as a prototype case of DNA bending in protein DNA complexes by Jen-Jacobson [191], who observed that in this and related structures, $\Delta H°$ and $\Delta S°$ are both positive whereas in complexes such as λ repressor-operator in which there is not a significant change in DNA bending on complexation, $\Delta H°$ and $\Delta S°$ are both negative. (In λ there are however other adaptive changes [209]. Methodology for calculating the configurational entropy change on protein-nucleic acid complexation on complexation has been successfully implemented and tested for λ and CAP (**Figure 9**). In the CAP protein DNA complex the DNA in the crystal form is nicked, which appears to have been necessary to get the complex to bind well enough for the crystal to form (subsequent un-nicked structures have been reported). This of course raises the issue of whether the nick influences the DNA structure and whether the results is realistic glimpse of CAP binding to intact DNA in solution, or to some extent an artifact of the nicked construct. Our MD on the complex (10 ns to date) shows the bending of intact DNA in the complex decrease from ca. 90° decreases to some 65°, still of course a significant deformation (**Figure 8a**). An experimental measurement of 77° for this angle has just been reported based on FRET [213] and 69° from topology measurements [214, 215], indicating our calculated trend to be consistent with the observed. A second issue is how much of the bending is intrinsic to the CAP DNA sequence, and how much is induced as a consequence of the complexation with protein. Consideration of this to date, including that of Jen-Jacobson, has been mainly in the context of protein induced bending. However the reference state for this idea is the canonical B form of random sequence DNA, since crystals of this length of oligonucleotide do not form in the absence of protein and it is beyond the present scope of NMR structure determination on free DNA. This length of sequence is now readily accessible to MD modeling, in which both the prediction of the solution structure of the sequence can be obtained and, by beginning a simulation with the bent form observed in the complex, the (reverse) time course of the adaptation process followed in detail. Preliminary results of this have been

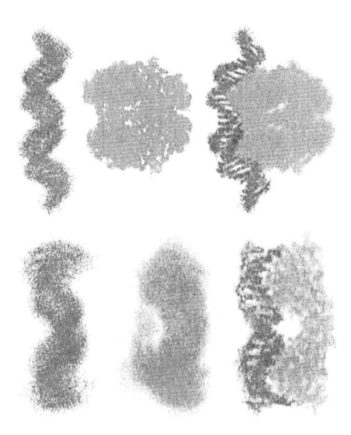

Figure 7. An overlay of superimposed structures from the MD simulation of CAP-DNA complex (top) and λ repressor-operator DNA complex (lower) and the free forms of the constituent proteins and DNA respectively. Different structures have their own characteristic dynamic behavior and this constitutes an important aspect of their recognition properties. MD simulations provide a unique opportunity to develop an understanding of these features responsible in the thermodynamics of recognition in protein-DNA complexes. Starting from the available experimentally solved structures, issues such as dynamical adaptation and the conformational capture hypothesis can be appreciated by performing *in silico* experiments to addresses various structural and energetic properties of the system.

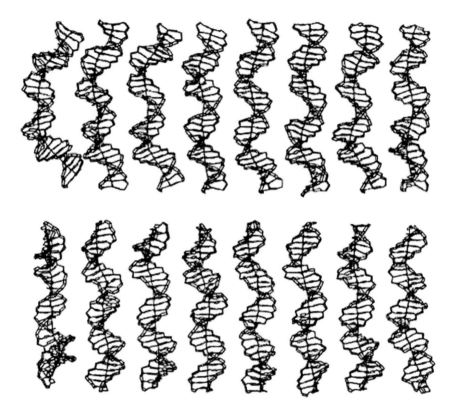

Figure 8a. hange in curvature of DNA during a 3.5 ns molecular dynamics of the curved DNA from the CAP-DNA complex (NDB id r023) in the absence of the protein as viewed from the two orthogonal directions at 500 ps intervals. The curved DNA in the original crystal with an approximate 90 bend induced by two major kinks diminishes as the DNA is relaxed in the absence of the protein.

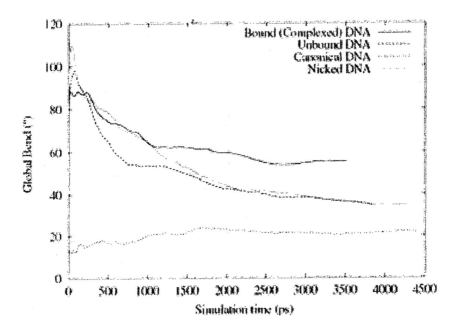

Figure 8b. he change in curvature if CAP binding DNA in various protein bound and unbound forms is shown. The bend DNA bound to the protein attains curvature of about 60 degrees during equilibration in solution. On the other hand, the DNA in absence of protein show a lower intrinsic bend of about 35 degrees. The canonical DNA attains a bend of about 25 degrees. It can be expected that from longer simulations and more sampling these two forms of DNA structure would achieve a common global bend angle, the intrinsic curvature of this DNA sequence. The rest of the observed DNA bend in the complex may be attributed to the protein induced effect.

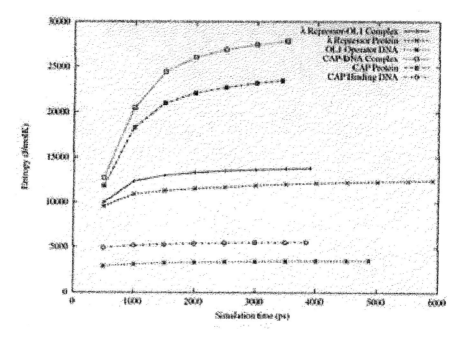

Figure 9. stimates of the configuration entropy based on quasi-harmonic analysis of MD trajectories using the Schlitter method [228]. The results indicate that structurally strained complexes such as the CAP-DNA system are entropically more favored by ~ 40 kcal/mol in comparison to system such as the λ repressor operator complex, which exhibits no significant DNA curvature. These results independently support the explanation advanced by Jen-Jacobson and co-workers [191].

obtained (**Figure 8b**) considering both the solution structure of the free unbound sequence and the complexed form of the DNA. The MD predicts the structural adaptation on CAP DNA complexation to be ~40% intrinsic and 60% protein induced.

In addition to λ repressor-operator and CAP, MD studies of two additional protein nucleic complexes are well along. The first involves the specificity of Papilloma virus E2 protein-DNA binding depends critically upon the sequence of a region of the DNA not in direct contact with the protein and, one of the simplest known examples of indirect readout [216, 217]. In order to study the role of structural adaptation of the DNA in the binding process, a subsequent simulation on the 16-mer cognate sequence d(AACCGAATTCGGTTG) was initiated from the crystallographic coordinates of the bound DNA in the crystal structure of the protein DNA complex [60]. MD simulations starting with the protein bound form relaxed rapidly back to the dynamical structure predicted from the previous simulations on the uncomplexed DNA. This indicates that the bound form E2 DNA is a dynamically unstable structure in the absence of protein, and the bound form is induced by the interaction with protein. In order to study the role of structural adaptation of the DNA in the binding process, a subsequent simulation on the 16-mer cognate sequence d(AACCGAATTCGGTTG) was initiated from the crystallographic coordinates of the bound DNA in the crystal structure of the protein DNA complex. MD simulations starting with the protein bound form relaxed rapidly back to the dynamical structure predicted from the previous simulations on the uncomplexed DNA. This indicates that the bound form E2 DNA is a dynamically unstable structure in the absence of protein, and the bound form is induced by the interaction with protein. The second is a problem in protein RNA complexation. Here leading evidence speaks for a special role for dynamical adaptation and the role of metastable states preorganized for binding [218], also known as the conformational capture hypothesis [202]. The role of solvent effects on these phenomena is important to clarify further at the molecular level. A detailed MD study of RNA protein complexation [212], yielded some provocative ideas about induced fit and conformational capture. In this system, the RNA adaptation involves an unstable form, whereas the protein adaptation utilizes a non-native conformational substate on binding, raising the provocative idea that molecular substates may code for specificity. Obtaining reliable treatments of the functional energetics and of protein-DNA complexation from MD is now an issue of considerable consequence.

Structural Bioinformatics of DNA: Regulatory Protein Binding and Site Recognition.

The proposal that protein DNA recognition involves not only the pattern of non bonded contacts presented by the DNA but the local molecular geometry of the DNA as well dates back at least twenty years [49]. The nonbonded contacts are well defined by base pair sequence, and thus sequence per se has been the basis of all practical bioinformatics tools aimed at locating regulatory regions via genomic scans [219]. Lafontaine and Lavery [220-222] have introduced an approach based on energy functions for identification of optimal interaction sites on DNA. Our own initiative aims at developing methods in which sequence and structure together or separately can systematically serve as a basis for recognizing protein DNA binding sites in genomic DNA. We have a basic idea and some promising preliminary studies on how to proceed with this [223]. However, much remains to be done to establish the viability of this method, to determine its corresponding range of applicability and explore the extent to which this method or any variations on the theme are the best way to proceed. The general idea opens up a number of new avenues of research, with implications both for genomics (an enhanced bioinformatics tool for identification of promoter sites) and molecular biophysics (the role of indirect readout in protein DNA recognition.)

The results we have to date on this project are described in a recent publication [223] and summarized schematically in **Figure 10**. A prototype two-parameter case of Boltzmann probability model of sequence dependent DNA structure has been derived from our current data base of all-atom MD simulations on oligonucleotides incorporated into Hidden Markov Models (HMMs) by a two step transformation/renormalization procedure applied to HMM emission probabilities, resulting in a prototype of a bioinformatics tool that should recognize molecular structural signals as well as sequence in protein-DNA binding sites on a genome. The binding of catabolite activator protein (CAP) to cognate DNA sequences was used as a prototype case for implementation and testing of the method. The results indicate that even HMMs based on probabilistic roll/tilt dinucleotide models of sequence dependent DNA structure have some capability to discriminate between known CAP binding and non-binding sites and to predict putative CAP binding sites in unknowns. Restricting HMMs to sequence only in regions of strong consensus in which the protein makes base specific contacts with the cognate DNA further improved the discriminatory capabilities of the HMMs. Comparison of results with controls based on sequence only indicates that extending the definition of consensus from sequence to structure improves the transferability of the HMMs. A preliminary result on scanning the E. coli genome has been carried out.

The method described is readily extended to definitions of sequence dependent DNA structure involving additional helicoidal parameters, and to include sequence context effects with trinucleotide or higher order models. The next round of questions we intend to address in this area are a) what is the most general formulation of the method, i.e. for n base pair steps described by m structural parameters b) can we provide even more convincing assessments and demonstrations that the method works as advertised? c) Is there a model problem for which we can know the answer that can serve as a basis for testing

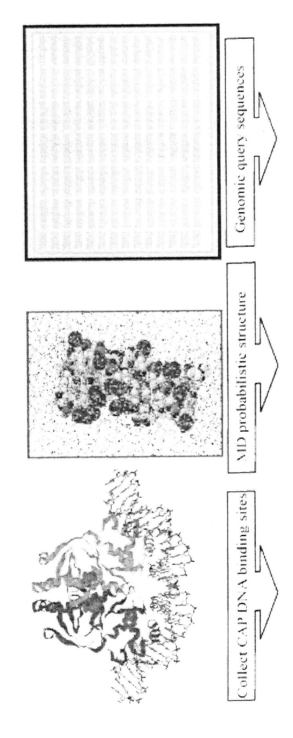

Genomic query sequences

MD probabilistic structure

Collect CAP DNA binding sites

46

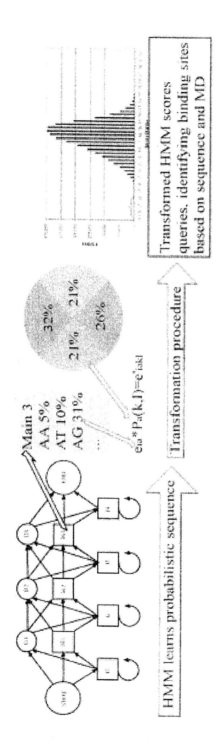

Figure 10. CAP binding site recognition based on Hidden Markov Models (HMM) including sequence and MD structure. CAP binding site data was collected from the literature and used to train HMM based on sequence. The HMM emission probabilities are transformed with description of the sequence dependent structure from MD, producing an HMM which 'knows' sequence and structure. The HMM can be used to score query sequences of interest [223].

the methodology further? d) Is it possible to take this problem in two stages, biochemical and molecular biological, i.e. separating the binding problem from the more complex promoter site recognition problem? e) How should one clean the data in order to proceed reliably with HMM development? Preliminary considerations of the promoter site recognition problem based on CAP demonstrates that the extensive molecular biology data on this system implies multiple binding sites [224], and thus one must develop well defined sub-classifications in which the presumption of a common mode of action is more secure. A final question we will pursue is f) What other possible approaches to this problem are there? With even leading answers to these questions, a viable line of development towards a robust, quantitative structural bioinformatics based on MD probability models of structure at the molecular level may be established.

The procedure by which we incorporate MD results into HMM treats sequence and structure parts as independent. The sequence HMM is trained by conventional methods against known binding data, which produces a set of HMM emission probabilities. An enhanced HMM based on MD is produced by a subsequent two-step transformation/renormalization process with the capability of recognizing both sequence and structure. A simple variation of this theme is to restrain the HMM to sequence only in consensus regions and apply the full sequence/structure in non-consensus regions of sequence. As noted in our preliminary results, a simple two parameter model of DNA structure by dinucleotide step has been reported which explores basic feasibility issues. However, our current operational implementation of this idea is a six parameter helicoidal model consisting of roll, tilt, twist, shift, slide and rise, coupled with considerations of which level of step model will serve. Dinucleotide and trinucleotide options fall out as a special case of this, together with the ability to assess the adequacy of the various reduced representations of the system. An alternative approach based on morphological indices of structure such as groove width and axis bending is attractive since it represents more what is seen by a protein in a recognition event, but is not as simple as it sounds when it comes to formulating a unique implementation. A viable alternative is sequence dependent two parameter binding/writhe approach derived from the normal vector method of Dickerson and coworkers and oligonucleotide persistence analysis from this laboratory [225]. This approach is isomorphic in methodology with our published formulation of the problem, can be readily reference to dinucleotide, tetranucleotide or other representations of the sequence effects. Furthermore, and a novel decomposition of intrinsic and induced effects may be applied, the latter of which fine tunes the model to a specific class of binding problem, which may well be the key step.

We have selected CAP DNA as our test case, this being one of the most extensively studied systems from the point of view of structural biology, biochemistry and molecular biology. An actual protein DNA regulatory event, the ultimate target of our structural bioinformatics, is a complex process the mechanism of which is not fully known. Thus as a stepping-stone, we propose to apply our methodology to thermodynamic data on CAP binding obtained in well controlled experiments at the biochemical level. We are also informed by general theoretical consideration on this problem by Stormo and coworkers[201, 226]. In this problem area, one is immediately confronted with the difficulty of forming a satisfactory data base of CAP binding sites to use for HMM training and testing. There is no shortage of CAP data available, but the extent to which putative sites have been established varies considerably. Three recognized classes of CAP sites imply different binding site positions and ways of interacting with the transcription machinery, as well as interactions with additional transcription factors that can be promoter specific [227]. This suggests more than one binding mechanism may need to be considered, and if so the data will need to be sub-classified accordingly in order to avoid mixed modalities at the HMM level. To proceed, the CAP literature has been examined and parsed for cases in which the actual the promoter sequence is available, the transcription start site(s) determined, a footprint in the promoter region has been determined, and in vivo functionality at activating or repressing a gene is well demonstrated. The transcription start site(s) must be known because CAP, other transcription factors, the transcription machinery, and other promoter elements are positioned with respect to this reference. The footprint is required to rule out promoters that are activated secondarily due to CAP manipulations. Sites that are proposed merely on the basis of functionality and sequence homology are not sufficient for our purposes. The construction of a database of all CAP binding sites taken from the literature to date parsed with respect to how well they fit the above criteria and classified based on binding mechanism is well underway. Entries include classifying CAP sites into the three classes based on sites of action on a genome and implication of other proteins. Entries include the above mentioned criteria plus the binding constant, the location with respect to the start site, 5' or 3' orientation designation, and a list of other known proteins in the promoter with their binding locations. Information will be easily extractable by user-specified search criteria.

Summary and Conclusions

The field of MD on DNA, facilitated by high performance computing, is now evolving rapidly. Methodological and simulation protocol issues are well appreciated and the remaining problems are each the subject of current investigations which will lead in the near future to the ready availability of MD models for nucleic acids relevant to structural and molecular biology. The next five years are likely to show considerable progress on the use of MD in describing the dynamical structure of DNA and understanding the structural and molecular biology of DNA and protein DNA interactions at the molecular level.

Acknowledgements.
This work builds on a series of projects carried out at Wesleyan University particularly by Dr. Matthew Young, Dr. Dennis Sprous, and Dr. Kevin McConnell. Financial support for this research was provided by NIGMS Grant # GM37909 (DLB, PI). Dr. Gabriela Barreiro acknowledges CNPq/Brazil (Conselho Nacional de Desenvolvimento Científico e Tecnológico) for a post-doctoral fellowship. The participation of Kelly M. Thayer in this project was supported by an NIGMS Training grant in Molecular Biophysics to Wesleyan University, Grant # GM 08271. Supercomputer time for our calculations was generously provided under auspices of PACI program on the facilities of the NCSA at the University of Illinois at Champaign/Urbana. DLB is pleased to acknowledge the kind hospitality of the Laboratoire de Biochemie Theorique of the Institut de Biochemie Physico-Chimique in Paris during a sabbatical in Spring 2002 and two month visit in Spring 2003 during which this article was completed.

References
1. Beveridge, D.L. and K.J. McConnell, *Nucleic acids: theory and computer simulation, Y2K.* Curr Opin Struct Biol, 2000. **10**(2): p. 182-196.
2. Beveridge, D.L., et al., *Molecular Dynamics Simulations on the Hydration, Structure and Motions of DNA Oligomers*, in *Water and Biological Molecules*, E. Westhof, Editor. 1993, The Macmillan Press, Ltd.: London. p. 165-225.
3. Beveridge, D.L. and G. Ravishanker, *Molecular Dynamics Studies of DNA.* Curr. Opin. Struct. Bio., 1994. **4**: p. 246-255.
4. Beveridge, D.L., et al., *Molecular Dynamics Simulations of DNA and Protein-DNA Complexes Including Solvent: Recent Progress.* ACS symposium series, 1994. **568**: p. 381-394.
5. Cheatham, T.E., 3rd and M.A. Young, *Molecular dynamics simulation of nucleic acids: Successes, limitations, and promise.* Biopolymers, 2001. **56**(4): p. 232-56.
6. MacKerell, A.D., Jr., N. Banavali, and N. Foloppe, *Development and current status of the CHARMM force field for nucleic acids.* Biopolymers, 2000. **56**(4): p. 257-65.
7. Beveridge, D.L., M.A. Young, and D. Sprous, *Modeling of DNA via Molecular Dynamics Simulation: Structure, Bending, and Conformational Transitions*, in *Molecular Modeling of Nucleic Acids*, N.B. Leontis and J. Santa Lucia, J., Editors. 1998, American Chemical Society: Washington, D.C. p. 260-284.
8. Beveridge, D.L. and R. Lavery, *Force Fields For Molecular Mechanics and Dynamics on DNA and RNA.* 2003. **MS in Preparation**.

50

9.   Jayaram, B. and D.L. Beveridge, *Modeling DNA in Aqueous Solution: Theoretical And Computer Simulation Studies on the Ion Atmosphere of DNA*. Ann. Rev. Biophys. Biomol. Struct., 1996. **25**: p. 367-394.

10.   Cheatham, T.E., 3rd and P.A. Kollman, *Molecular dynamics simulation of nucleic acids*. Annu Rev Phys Chem, 2000. **51**: p. 435-71.

11.   Giudice, E. and R. Lavery, *Simulations of nucleic acids and their complexes*. Acc Chem Res, 2002. **35**(6): p. 350-7.

12.   Norberg, J. and L. Nilsson, *Molecular dynamics applied to nucleic acids*. Acc Chem Res, 2002. **35**(6): p. 465-72.

13.   Orozco, M., et al., *Theoretical Methods for the Simulation of Nucleic Acids*. Chem. Soc. Rev., 2003. **32**(6): p. 350-364.

14.   Darden, T., D. York, and L. Pedersen, *The Particle Mesh Ewald Method*. J. Chem. Phys., 1995. **98**: p. 10089-10092.

15.   Smith, P.E. and B.M. Pettitt, *Ewald Artifacts in Liquid State Molecular Dynamics Simulations*. J. Chem. Phys., 1996. **105**(10): p. 4289.

16.   Cornell, W.D., et al., *A Second Generation Force Field for the Simulation of Proteins, Nucleic Acids, and Organic Molecules*. J. Am. Chem. Soc., 1995. **117**(19): p. 5179-5197.

17.   Cheatham, T.E., et al., *Molecular dynamics simulations on nucleic acid systems using the Cornell et al force field and particle mesh Ewald electrostatics*, in *Molecular Modeling of Nucleic Acids*. 1998, Amer Chemical Soc: Washington. p. 285-303.

18.   Young, M.A., G. Ravishanker, and D.L. Beveridge, *A 5-Nanosecond Molecular Dynamics Trajectory for B-DNA: Analysis of Structure, Motions and Solvation*. Biophys. J., 1997. **73**(5): p. 2313-2336.

19.   Young, M.A., B. Jayaram, and D.L. Beveridge, *Intrusion of Counterions into the Spine of Hydration in the Minor Groove of B-DNA: Fractional Occupancy of Electronegative Pockets*. J. Am. Chem. Soc., 1997. **119**(1): p. 59-69.

20.   Kabsch, W., C. Sander, and E.N. Trifonov, *The Ten Helical Twist Angles of B-DNA*. Nucleic Acids Res., 1982. **10**(3): p. 1097-1104.

21.   Cheatham, T.E., 3rd, P. Cieplak, and P.A. Kollman, *A modified version of the Cornell et al. force field with improved sugar pucker phases and helical repeat*. J Biomol Struct Dyn, 1999. **16**(4): p. 845-62.

22.   Foloppe, N. and J. MacKerell, A.D., *All-atom empirical force field for nucleic acids: I. Parameter optimization based on small molecule and condensed phase macromolecular target data*. Journal of Computational Chemistry, 2000. **21**(2): p. 86-104.

23.   MacKerell, J., A.D. and N. Banavali, *All-atom empirical force field for nucleic acids: II. Application to molecular dynamics simulations of DNA and RNA in solution*. Journal of Computational Chemistry, 2000. **21**(2): p. 105-120.

24.   Langley, D.R., *The BMS nucleic acid force field*. 1996, Bristol-Myers Squibb Company, Wallingford, CT.

25.   Langley, D.R., *Molecular dynamic simulations of environment and sequence dependent DNA conformations: the development of the BMS nucleic acid force field and comparison with experimental results.* J. Biomol. Struct. Dyn., 1998. **16**(3): p. 487-509.

26.   Feig, M. and B.M. Pettitt, *Structural equilibrium of DNA represented with different force fields.* Biophys. J., 1998. **75**(1): p. 134-149.

27.   Reddy, S.Y., F. Leclerc, and M. Karplus, *DNA polymorphism: a comparison of force fields for nucleic acids.* Biophys J, 2003. **84**(3): p. 1421-49.

28.   Haile, J.M., *Molecular Dynamics Simulation: Elementary Methods.* 1992, New York: John Wiley and Sons, Inc.

29.   Poncin, M., B. Hartmann, and R. Lavery, *Conformational sub-states in B-DNA.* J. Mol. Biol., 1992. **226**(3): p. 775-794.

30.   McConnell, K.M., et al., *A Nanosecond Molecular Dynamics Trajectory for a B DNA Double Helix: Evidence for Substates.* J. Am. Chem. Soc., 1994. **116**: p. 4461-4462.

31.   Varnai, P., et al., *Alpha/gamma transitions in the B-DNA backbone.* Nucleic Acids Res, 2002. **30**(24): p. 5398-406.

32.   Djuranovic, D. and B. Hartmann, *Conformational Characteristics and Correlations in Crystal Structures of Nucleic Acid Oligonucleotides.* J. Biomol. Struct. Dyn., 2003. **20**: p. 1-17.

33.   Judson, H.F., *The Eighth Day of Creation.* 1979, New York: Simon and Shuster, Inc.

34.   Saenger, W., *Principles of Nucleic Acid Structure.* 1984, New York: Springer Verlag.

35.   Sinden, R.R., *DNA Structure and Function.* 1994, San Diego; London: Academic Press. 398.

36.   Calladine, C.R. and H.R. Drew, *Understanding DNA: The Molecule and How it Works.* 1997, San Diego, CA: Academic Press.

37.   Neidle, S., ed. *Oxford handbook of nucleic acid structure.* 1999, Oxford University Press: Oxford, New York. 662.

38.   Lavery, R. and K. Zakrzewska, *Base and base pair morphologies, helical parameters, and definitions,* in *Oxford handbook of nucleic acid structure,* S. Neidle, Editor. 1999, Oxford University Press: Oxford, New York. p. 39-76.

39.   Arnott, S., P.J. Campbell-Smith, and R. Chandrasekaran, *Atomic Coordinates and Molecular Conformations for DNA-DNA, RNA-RNA, and DNA-RNA Helices,* in *CRC Handbook of Biochemistry and Molecular Biology,* G. Fasman, Editor. 1976, CRC Press: Cleveland. p. 411-422.

40.   Cheatham III, T.E., et al., *Molecular Dynamics and Continuum Solvent Studies of the Stability of PolyG-PolyC and PolyA-PolyT DNA Duplexes in Solution.* J. Biomol. Struct. Dyn., 1998. **16**(2): p. 265-280.

41.   Steitz, T.A., *Structural Studies of Protein-Nucleic Acid Interactions: The Sources of Sequence-Specific Binding*. Quart. Rev. Biophys., 1990. **23**: p. 205-280.

42.   Travers, A., *DNA-Protein Interactions*. 1993, London: Chapman and Hall.

43.   Arnott, S. and E. Selsing, *Structures for the polynucleotide complexes Poly(dA)-Poly(dT) and Poly(dA)-Poly(dT)-Poly(dT)*. Journal of Molecular Biology, 1974. **88**: p. 509-521.

44.   Rosenberg, J.M., et al., *Double Helix at Atomic Resolution*. Nature, 1973. **243**( ): p. 150-154.

45.   Wing, R.M., et al., *Crystal Structure Analysis of a Complete Turn of B-DNA*. Nature, 1980. **287**: p. 755-758.

46.   Drew, H.R., et al., *Structure of a B DNA Dodecamer I: Conformation and Dynamics*. Proc. Nat. Acad. Sci. USA, 1981. **78**: p. 2179-2183.

47.   McClarin, J.A., et al., *Structure of the DNA- Eco RI Endonuclease Recognition Complex at 3Å Resolution*. Science, 1986. **234**: p. 1526-1540.

48.   Kopka, M.L., et al., *The Binding of Netropsin to Double Helical DNA of Sequence CGCGAATTCGCG: Single Crystal X-ray Structure Analysis*, in *Structure and Motion: Membranes, Nucleic Acids and Proteins*, E. Clementi, et al., Editors. 1985, Adenine Press: Guilderland, N.Y. p. 461ff.

49.   Dickerson, R.E., *The DNA Helix and How It Is Read*. Sci. Am., 1983. **249**: p. 94-111.

50.   Drew, H.R. and R.E. Dickerson, *Structure of a B-DNA Dodecamer III. Geometry of Hydration*. J. Mol. Biol., 1981. **151**: p. 535-556.

51.   Dickerson, R.E., H.R. Drew, and B. Connor, *Single Crystal X-ray Structure Analysis of A, B and Z Helices or One Good Turn Deserves Another*, in *Biomolecular Stereodynamics*, R.H. Sarma, Editor. 1981, Adenine Press: Guilderland, NY.

52.   Berman, H.M., et al., *The Nucleic Acid Database: A Comprehensive Relational Database of Three-dimensional Structures of Nucleic Acids*. Biophys. J., 1992. **63**: p. 751-759.

53.   Berman, H.M., et al., *The Nucleic Acid Database Project*, in *Biological Structures and Dynamics*, R.H. Sarma and M.H. Sarma, Editors. 1996, Adenine Press: Albany. p. 1-13.

54.   Dickerson, R.E., D.S. Goodsell, and S. Neidle, *...the tyranny of the lattice...* Proc. Natl. Acad. Sci. (USA), 1994. **911**: p. 3579-3583.

55.   DiGabriele, A.D., M.R. Sanderson, and T.A. Steitz, *Crystal Lattice Packing is Important in Determining the Bend of a DNA Dodecamer Containing an Adenine Tract*. Proceedings of the National Academy of Sciences (USA), 1989. **86**: p. 1816-1820.

56.   Hizver, J., et al., *DNA bending by an adenine-thymine tract and its role in gene regulation*. Proc Natl Acad Sci U S A, 2001. **98**(15): p. 8490-5.

57.   Sprous, D., et al., *Dehydrating Agents Sharply Reduce Curvature in DNAs Containing A-Tracts.* Nucleic Acids Research, 1995. **23**: p. 1816-1821.

58.   Lee, H., T. Darden, and L. Pederson, *Accurate Crystal Molecular Dynamics Simulations using Particle Mesh Ewald: RNA Dinucleotides: ApU and GpC.* Chem. Phys. Lett., 1995. **243**: p. 229-235.

59.   Bevan, D.R., et al., *Molecular Dynamics Simulations of the d(CCAACGTTGG)2 Decamer: Influence of the Crystal Environment.* Biophysical Journal, 2000. **78**(2): p. 668-682.

60.   Byun, S.K. and D.L. Beveridge, *Molecular Dynamics Simulations Of Papilloma Virus E2 DNA Sequences: Dynamical Models for Oligonucleotide Structures in Solution.* Biopolymers, 2003. **In Press.**

61.   Wuthrich, K., *Nmr - This Other Method for Protein and Nucleic-Acid Structure Determination.* Acta Crystallographica Section D-Biological Crystallography, 1995. **51**: p. 249-270.

62.   Tjandra, N., et al., *The NMR Structure of a DNA Dodecamer in an Aqueous Dilute Liquid Crystalline Phase.* J. Am. Chem. Soc., 2000. **122**(26): p. 6190-6200.

63.   MacDonald, D. and P. Lu, *Residual Dipolar Couplings in Nucleic Acid Structure Determination.* Curr. Op. Struct. Biol., 2002. **12**: p. 337-343.

64.   Vermulen, A., H. Zhou, and A. Pardi, *Determining DNA Global Structure and DNA Bending by Application of NMR Residual Dipolar Couplings.* J. Am. Chem. Soc., 2000. **122**: p. 9638-9647.

65.   Trantfrek, L., et al., *A Method for Direct Determination of Helical PArameters in Nucleic Acids Using Residual Dipolar Couplings.* J. Am. Chem. Soc., 2000. **122**: p. 10454-10455.

66.   Arthanari, H., et al., *Assessment of the molecular dynamics structure of DNA in solution based on calculated and observed NMR NOESY volumes and dihedral angles from scalar coupling constants.* Biopolymers, 2003. **68**(1): p. 3-15.

67.   Pitici, F., S.B. Dixit, and D.L. Beveridge, *Comparisions of the Structure of the d(GGCAAAAAACGG) duplex from Molecular Dynamics Simulations, NMR Spectroscopy and Crystallography.* MS in Prep., 2003.

68.   Falk, M., J. Hartman, K. A. , and R.C. Lord, *Hydration of Deoxyribnucleic Acid I. A Gravimetric Study.* Journal of the American Chemical Society, 1962. **84**: p. 3843-3846.

69.   Falk, M., J. K.A. Hartmann, and R.C. Lord, *Hydration of Deoxyribnucleic Acid III. A Spectroscopic Study of the Effect of Hydration on the Structure of Deoxyribonucleic Acid.* Journal of the American Chemical Society, 1963. **85**: p. 391-394.

70.   Falk, M., J. K. A. Hartman, and R.C. Lord, *Hydration of Deoxyribnucleic Acid II. An Infrared Study.* Journal of the American Chemical Society, 1963. **85**: p. 387-391.

71.   Falk, M., A.G. Poole, and C.G. Goyman, *IR Study of the State of Water in the Hydration Shell of DNA.* Can. J. Chem., 1970. **48**: p. 1536-1542.

72.   Westhof, E. and D.L. Beveridge, *Hydration of Nucleic Acids,* in *Water Science Reviews: The Molecules of Life,* F. Franks, Editor. 1989, Cambridge University Press: Cambridge. p. 24-136.

73.   Saenger, W., W.N. Hunter, and O. Kennard, *DNA conformation is determined by economics in the hydration of phosphate groups.* Nature, 1986. **324**(6095): p. 385-388.

74.   Olson, W.K. and V.B. Zhurkin, *Twenty Years of DNA Bending,* in *Biological Structure and Dynamics,* R.H. Sarma and M.H. Sarma, Editors. 1996, Adenine Press: Albany, NY. p. 341-370.

75.   Pettitt, B.M., V.A. Makarov, and B.K. Andrews, *Protein hydration density: theory, simulations and crystallography.* Current Opinion in Structural Biology, 1998. **8**(2): p. 218-221.

76.   Schneider, B., K. Patel, and H.M. Berman, *Hydration of the phosphate group in double-helical DNA.* Biophysical Journal, 1998. **75**(5): p. 2422-2434.

77.   Schneider, B. and H.M. Berman, *Hydration of the DNA bases is local.* Biophys. J, 1995. **69**(6): p. 2661-9.

78.   Beveridge, D.L., et al., *Aqueous hydration of nucleic acid constituents: Monte Carlo computer simulation studies.* J. Biomol. Struct. Dyn., 1984. **2**(2): p. 261-270.

79.   Young, M.A., B. Jayaram, and D.L. Beveridge, *Local Dielectric Environment of B-DNA in Solution: Results from a 14 Nanosecond Molecular Dynamics Trajectory.* The Journal of Physical Chemistry B, 1998. **102**(39): p. 7666-7669.

80.   Makarov, V.A., et al., *Diffusion of solvent around biomolecular solutes: a molecular dynamics simulation study.* Biophys. J., 1998. **75**(1): p. 150-158.

81.   Feig, M. and B.M. Pettitt, *Modeling High-resolution Hydration Patterns in Correlation with DNA Sequence and Conformation.* J. Mol. Biol., 1999. **286**(4): p. 1075-1095.

82.   Pohl, F.M. and T.M. Jovin, *Salt-Induced Cooperative Conformational Change of a Synthetic DNA: Equilibrium and Kinetic Studies with poly d(G-C).* Journal of Molecular Biology, 1972. **57**: p. 373-395.

83.   Manning, G.S., *The Molecular Theory of Polyelectrolyte Solutions with Applications to the Electrostatic Properties of Polynucleotides.* Quart. Rev. Biophys., 1978. **11**: p. 179-246.

84.   Record Jr., M.T., C.F. Anderson, and T.M. Lohman, *Thermodynamics Analysis of Ion Effects on the Binding and Conformational Equilibria of Proteins and Nucleic Acids: the Role of Ion Association and Release, Screening, and Ion Effects on Water Activity.* Quart. Rev. Biophys., 1978. **11**: p. 103-178.

85. LeBret, M. and B.H. Zimm, *Distribution of Counterions around a Cylindrical Polyelectrolyte and Manning's Condensation Theory.* Biopolymers, 1984. **23**: p. 287-312.

86. Mills, P., C.F. Anderson, and M.T. Record Jr., *Monte Carlo Studies of Counterion-DNA Interactions. Comparison of the Radial Distribution of Counterions with Predictions of Other Polyelectrolyte Theories.* J. Phys. Chem., 1985. **89**: p. 3984-3994.

87. Mills, P., C.F. Anderson, and M.T. Record Jr., *Grand Canonical Monte Carlo Calculations of Thermodynamic Coefficients for a Primitive Model of DNA-Salt Solutions.* J. Phys. Chem., 1986. **90**: p. 6541-6548.

88. Bacquet, R. and P.J. Rossky, *Ionic Atmosphere of Rodlike Polyelectrolytes. A Hypernetted Chain Study.* J. Phys. Chem., 1984. **88**: p. 2660-2669.

89. Bleam, M.L., C.F. Anderson, and M.T. Record Jr., *Relative Binding Affinities of Monovalent Cations for Double Stranded DNA Studied by Sodium-23 NMR.* Proceedings of the National Academy of Sciences (USA), 1980. **77**: p. 3085-3089.

90. Bleam, M.L., C.F. Anderson, and M.T. Record Jr., *Sodium-23 Nuclear Magnetic Resonance Studies of Cation-DNA Interactions.* Biochemistry, 1983. **22**: p. 5418-5425.

91. Record Jr., M.T., et al., *Ions as Regulators of Protein-Nucleic Acid Interactions In Vitro and In Vivo.* Adv. Biophys., 1985. **20**: p. 109 - 135.

92. Feig, M. and B.M. Pettitt, *Sodium and chlorine ions as part of the DNA solvation shell.* Biophys. J., 1999. **77**(4): p. 1769-1781.

93. Lavery, R. and B. Pullman, *The molecular electrostatic potential, steric accessibility and hydration of Dickerson's B-DNA dodecamer d(CpGpCpGpApApTpTpCpGpCpG).* Nucleic Acids Research, 1981. **9**(15): p. 3765-77.

94. Rosenberg, J.M., et al., *Double helix at atomic resolution.* Nature, 1973. **243**(5403): p. 150-4.

95. Hud, N.V. and J. Feigon, *Localization of Divalent Metal Ions in the Minor Groove of DNA A-tracts.* J. Am. Chem. Soc., 1997. **119**: p. 5756-5757.

96. Hud, N.V., V. Sklenar, and J. Feigon, *Localization of ammonium Ions in the minor groove of DNA duplexes in solution and the origin of DNA A-tract bending.* Journal of Molecular Biology, 1999. **286**(3): p. 651-660.

97. Shui, X., et al., *Structure of the Potassium Form of CGCGAATTCGCG: DNA Deformation by Electrostatic Collapse around Inorganic Cations.* Biochemistry, 1998. **37**: p. 16877-16887.

98. Tereshko, V., G. Minasov, and M. Egli, *A "Hydrat-ion" Spine in a B-DNA Minor Groove.* J. Am. Chem. Soc., 1999. **121**(15): p. 3590-3595.

99. Lyubartsev, A.P. and A. Laaksonen, *Molecular dynamics simulations of DNA in solution with different counter-ions.* J. Biomol. Struct. Dyn., 1998. **16**(3): p. 579-592.

100. van Dam, L., et al., *Self-Diffusion and Association of Li+, Cs+, and H2O in Oriented DNA Fibers. An NMR and MD Simulation Study.* J. Phys. Chem. B, 1998. **102**(51): p. 10636-10642.

101. Rosenberg, J.M., et al., *RNA Double-Helical Fragments at Atomic Resolution: II. The Crystal Structure of Sodium Guanylyl -3',5'-Cytidine Nonhydrate.* Journal of Molecular Biology, 1976. **104**: p. 145-167.

102. Bonvin, A.M., *Localisation and dynamics of sodium counterions around DNA in solution from molecular dynamics simulation.* Eur Biophys J, 2000. **29**(1): p. 57-60.

103. McFail-Isom, L., C.C. Sines, and L.D. Williams, *DNA structure: cations in charge?* Current Opinion In Structural Biology, 1999. **9**(3): p. 298-304.

104. Hud, N.V. and M. Polak, *DNA-cation interactions: The major and minor grooves are flexible ionophores.* Curr Opin Struct Biol, 2001. **11**(3): p. 293-301.

105. Strauss, J.K. and L.J.r. Maher, *DNA bending by asymmetric phosphate neutralization.* Science, 1994. **266**(5192): p. 1829-34.

106. Chiu, T.K., M. Kaczor-Grzeskowiak, and R.E. Dickerson, *Absence of minor groove monovalent cations in the crosslinked dodecamer C-G-C-G-A-A-T-T-C-G-C-G.* Journal of Molecular Biology, 1999. **292**(3): p. 589-608.

107. Stellwagen, N.C., et al., *Preferential Counterion Binding to A-Tract DNA Oligomers.* J Mol Biol, 2001. **305**(5): p. 1025-1033.

108. Hamelberg, D., L.D. Williams, and W.D. Wilson, *Influence of the dynamic positions of cations on the structure of the DNA minor groove: sequence-dependent effects.* J Am Chem Soc, 2001. **123**(32): p. 7745-55.

109. Ponomarev, S., K.M. Thayer, and D.L. Beveridge, *Convergence of mobile Na+ ions after 60 ns of molecular dynamics on d(CGCGAATTCGCG).* J. Am. Chem. Soc., 2004. **Submitted.**

110. Bloomfield, V.A., *Condensation of DNA by multivalent cations: considerations on mechanism.* Biopolymers, 1991. **31**(13): p. 1471-81.

111. Boon, E.M. and J.K. Barton, *Charge transport in DNA.* Curr Opin Struct Biol, 2002. **12**(3): p. 320-9.

112. Aqvist, J., *Ion-Water Interaction Potentials Derived from Free Energy Perturbation Simulations.* J. Phys. Chem., 1990. **94**: p. 8021-8024.

113. Roux, B. and S. Berneche, *On the potential functions used in molecular dynamics simulations of ion channels.* Biophys J, 2002. **82**(3): p. 1681-4.

114. Halgren, T.A. and W. Damm, *Polarizable force fields.* Curr Opin Struct Biol, 2001. **11**(2): p. 236-42.

115. Dickerson, R.E., et al., *Definitions and Nomenclature of Nucleic Acid Structural Parameters.* EMBO J., 1989. **8**: p. 1-4.

116. Suzuki, M., et al., *Use of a 3D structure data base for understanding sequence-dependent conformational aspects of DNA.* J. Mol. Biol., 1997. **274**(3): p. 421-435.

117. El Hassan, M.A., *Conformational Characteristics of DNA: Empirical Classifications and a Hypothesis for the Conformational Behavior of Dinucleotide Steps.* Phil. Trans. Roy. Soc. London A, 1997. **355**: p. 43-100.

118. Liu, Y. and D.L. Beveridge, *A Refined Prediction Method for Gel Retardation of DNA Oligonucleotides from Dinucleotide Step Parameters:Reconciliation of DNA Bending Models with Crystal Structure Data.* J. Biomol. Strut. & Dyn., 2001. **18**: p. 505-526.

119. Olson, W.K., et al., *DNA sequence-dependent deformability deduced from protein-DNA crystal complexes.* Proc Natl Acad Sci U S A, 1998. **95**(19): p. 11163-8.

120. Nelson, C.M.H., et al., *The Structure of an Oligo(dA)oligo(dT) Tract and its Biological Implications.* Nature, 1987. **330**: p. 221-226.

121. Chan, S.S., et al., *Temperature-dependent Ultraviolet Resonance Raman Spectroscopy of the Premelting State of dAdT DNA.* Biophysical Journal, 1997. **72**(4): p. 1512-1520.

122. McConnell, K.J. and D.L. Beveridge, *Molecular dynamics simulations of B '-DNA: sequence effects on A-tract-induced bending and flexibility.* J Mol Biol, 2001. **314**(1): p. 23-40.

123. Sherer, E.C., et al., *Molecular Dynamics Studies of DNA A-Tract Structure and Flexibility.* J. Am. Chem. Soc., 1999. **121**(25): p. 5981-5991.

124. Hunter, C.A., *Sequence-Dependent DNA Structure: The Role of Base Stacking Interactions.* Journal of Molecular Biology, 1993. **1993**: p. 1025-1054.

125. Friedman, R.A. and B. Honig, *The Electrostatic Contribution to DNA Base-Stacking Interactions.* Biopolymers, 1992. **32**(2): p. 145-159.

126. Elcock, A.H. and J.A. McCammon, *Sequence Dependent Hydration of DNA: Theoretical Results.* J. Am. Chem. Soc., 1995. **117**: p. 10161-10162.

127. Lavery, R. and H. Sklenar, *The Definition of Generalized Helicoidal Parameters and of Axis Curvature for Irregular Nucleic Acids.* J. Biomol. Struct. Dyn., 1988. **6**: p. 63-91.

128. Lu, X.-J. and W.K. Olson, *Resolving the Discrepancies Among Nucleic Acid Conformational Analyses.* J. Mol. Biol., 1999. **285**: p. 1563-1575.

129. Dickerson, R.E., *Sequence-Dependent B-DNA Conformation in Crystals and in Protein Complexes.* Proceedings of the Tenth Conversation, State University of New York, Albany, 1997: p. 17-36.

130. Packer, M.J., M.P. Dauncey, and C.A. Hunter, *Sequence-dependent DNA Structure: Dinucleotide Conformational Maps.* J. Mol Biol., 2000. **295**(1): p. 71-83.

131. Packer, M.J., M.P. Dauncey, and C.A. Hunter, *Sequence-dependent DNA Structure: Tetranucleotide Conformational Maps.* J. Mol. Biol., 2000. **295**(1): p. 85-103.

132. Beveridge, D.L., et al., *Molecular Dynamics Simulations of the 136 Unique Tetranucleotide Sequences of DNA Oligonucleotides. I. Research Design, Informatics, and Preliminary Results.* MS. In Preparation, 2003.

133. Franklin, R.E. and R.G. Gosling, *The Structure of Sodium Thymonucleate Fibers I. The Influence of Water Content.* Acta Cryst., 1953. **6**: p. 673-677.

134. Wolf, B. and S. Hanlon, *Structural transitions of deoxyribonucleic acid in aqueous electrolyte solutions. II. The role of hydration.* Biochemistry, 1975. **14**(8): p. 1661-70.

135. Ivanov, V.I., et al., *The B to A Transition of DNA in Solution.* J. Mol. Biol., 1974. **87**: p. 817-833.

136. Pilet, J. and J. Brahms, *Dependence of B-A Conformational Change on Base Composition.* Nature New Biology, 1972. **236**: p. 99-100.

137. Pohl, F.M., *Polymorhism of a Synthetic DNA in Solution.* Nature, 1976. **260**: p. 365-366.

138. Nishimura, Y., C. Torigoe, and M. Tsuboi, *Salt Induced B-A transition of poly(dG)*poly(dC) and the stabilization of A-form by its Methylation.* Nucleic Acids Research, 1986. **14**: p. 2737-2748.

139. Cheatham III, T.E. and P.A. Kollman, *Observation of the A-DNA to B-DNA transition during unrestrained molecular dynamics in aqueous solution.* J. Mol. Biol., 1996. **259**(3): p. 434-444.

140. Sprous, D., M.A. Young, and D.L. Beveridge, *Molecular Dynamics Studies of the Conformational Preferences of a DNA Double Helix in Water and in an Ethanol/Water Mixture: Theoretical Considerations of the A/B Transition.* J. Phys. Chem., 1998. **102**: p. 4658-4667.

141. Cheatham, T.E. and P.A. Kollman, *Molecular dynamics simulation of nucleic acids in solution: how sensitive are the results to small perturbations in the force field and environment?* Struct., Motion, Interact. Expression Biol. Macromol., Proc. Conversation Discip. Biomol. Stereodyn., 10th, 1998. **1**: p. 99-116.

142. Soliva, R., et al., *Role of sugar re-puckering in the transition of A and B forms of DNA in solution. a molecular dynamics study.* J. Biomol. Struct. Dyn., 1999. **17**(1): p. 89-99.

143. Yang, Y. and B.M. Pettitt, *B to A Transition of DNA on the Nanosecond timescale.* J. Phys. Chem., 1996. **100**: p. 2564-2566.

144. Jayaram, B., et al., *Free Energy Analysis of the Conformational Preferences of A and B forms of DNA in Solution.* Journal of the American Chemical Society, 1998. **120**(41): p. 10629-10633.

145. Breslauer, K.J., *A Thermodynamic Perspective of DNA Bending.* Curr. Opin. Struct. Bio., 1991. **1**: p. 416-422.

146. Marini, J.C., et al., *Bent Helical Structure in Kinetoplast DNA.* Proceedings of the National Academy of Sciences (USA), 1982. **79**: p. 7664-7668.

147. Marini, J.C., et al., *Physical Characterization of a Kinetoplast DNA Fragment with Unusual Properties.* J. Biol. Chem., 1984. **259**: p. 8974-8979.

148. Ulanovsky, L., et al., *Curved DNA: Design, synthesis, and circularization.* Proc. Natl. Acad. Sci. USA, 1986. **83**: p. 862-866.

149. Griffith, J.D., et al. *Use of Electron Microscopy to Examine Sequence-Directed DNA Bending.* in *Structure and Expression: DNA bending and curvature.* 1988. State University of New York at Albany.

150. Hagerman, P.J., *Evidence for the existence of stable curvature of DNA in solution.* Proc. Natl. Acad. Sci, 1984. **81**: p. 4632-4636.

151. Parkhurst, K.M., M. Brenowitz, and L.J. Parkhurst, *Simultaneous binding and bending of promoter DNA by the TATA binding protein: real time kinetic measurements.* Biochemistry, 1996. **35**(23): p. 7459-65.

152. Klostermeier, D. and D.P. Millar, *Time-resolved fluorescence resonance energy transfer: a versatile tool for the analysis of nucleic acids.* Biopolymers, 2001. **61**(3): p. 159-79.

153. Wojtuszewski, K. and I. Mukerji, *HU binding to bent DNA: a fluorescence resonance energy transfer and anisotropy study.* Biochemistry, 2003. **42**(10): p. 3096-104.

154. Dlakic, M., et al., *The organic crystallizing agent 2-methyl-2,4-pentanediol reduces DNA curvature by means of structural changes in A-tracts.* J. Biol. Chem., 1996. **271**: p. 17911-17919.

155. Vacano, E. and P.J. Hagerman, *Analysis of birefringence decay profiles for nucleic acid helices possessing bends: the tau-ratio approach.* Biophys J, 1997. **73**(1): p. 306-17.

156. Widom, J., *Role of DNA sequence in nucleosome stability and dynamics.* Q Rev Biophys, 2001. **34**(3): p. 269-324.

157. Neidle, S., *Nucleic Acid Structure and Recognition.* 2002, Oxford UK: Oxford Press. 646.

158. Luger, K., et al., *Crystal structure of the nucleosome core particle at 2.8 A resolution.* Nature, 1997. **389**(6648): p. 251-60.

159. Ioshikhes, I., A. Bolshoy, and E.N. Trifonov, *Preferred positions of AA and TT dinucleotides in aligned nucleosomal DNA sequences.* Journal of Biomolecular Structure and Dynamics, 1992. **9**(6): p. 1111-7.

160. Hagerman, P.J., *Sequence-Directed Curvature of DNA.* Annu. Rev. Biochem., 1990. **59**: p. 755-781.

161. Bolshoy, A., et al., *Curved DNA without A-A: experimental estimation of all 16 DNA wedge angles.* Proc. Natl. Acad. Sci. U. S. A., 1991. **88**(6): p. 2312-6.

162. Narayana, N., et al., *Crystal and molecular structure of a DNA fragment: d(CGTGAATTCACG).* Biochemistry, 1991. **30**(18): p. 4449-55.

163. Dickerson, R.E., et al., *The Effect of Crystal Packing on Oligonucleotide Double Helix Structure.* J. Biomol. Struct. Dyn., 1987. **5**(3): p. 557-579.

164. Dickerson, R.E., et al., *Polymophism, Packing, Resolution, and Reliablity in Single-crystal DNA Oligomer Analyses.* Nucleosides & Nucleotides, 1991. **10**: p. 3-24.

165. Olson, W.K., et al., *Influence of Fluctuations on DNA Curvature: A Comparison of Flexible and Static Wedge Models of Intrinsically Bent DNA.* Journal of Molecular Biology, 1993. **232**(2): p. 530-554.

166. Goodsell, D.S. and R.E. Dickerson, *Bending and curvature calculations in B-DNA.* Nucleic Acids Research, 1994. **22**(24): p. 5497-503.

167. Trifonov, E.N., *Sequence-dependent Deformational Anisotropy of Chromatin DNA.* Nucleic Acids Research, 1980. **8**(17): p. 4041-4053.

168. Dickerson, R.E., D. Goodsell, and M.L. Kopka, *MPD and DNA bending in crystals and in solution.* J Mol Biol, 1996. **256**(1): p. 108-25.

169. Rouzina, I. and V.A. Bloomfield, *DNA Bending by Small Mobile, Multivalent Cations.* Biophys. J., 1998. **74**(6): p. 3152-3164.

170. Hud, N.V. and J. Plavec, *A unified model for the origin of DNA sequence-directed curvature.* Biopolymers, 2003. **69**(1): p. 144-58.

171. Young, M.A. and D.L. Beveridge, *Molecular Dynamics Simulations of an Oligonucleotide Duplex with Adenine Tracts Phased by a Full Helix Turn.* Journal of Molecular Biology, 1998. **281**(4): p. 675-687.

172. Sprous, D., M.A. Young, and D.L. Beveridge, *Molecular Dynamics Studies of Axis Bending in d(G5-(GA4T4C)2-C5) and d(G5-(GT4A4C)2-C5): Effects of Sequence Polarity on DNA Curvature.* J. Mol. Biol., 1999. **285**: p. 1623-1632.

173. Barreiro, G., S. Dixit, and D.L. Beveridge, *Molecular Dynamics Study of the Premelting Transition in a DNA Oligonucleotide with Phased A-tracts.* J. Am. Chem. Soc., 2004. **Submitted.**

174. Zhurkin, V.B., et al., *Static and statistical bending of DNA evaluated by Monte Carlo simulations.* Proc. Natl. Acad. Sci., 1991. **88**: p. 7046-7050.

175. Calladine, C.R., H.R. Drew, and M.J. McCall, *The Intrinsic Curvature of DNA in Solution.* Journal of Molecular Biology, 1988. **201**: p. 127-137.

176. Maroun, R.C. and W.K. Olson, *Base Sequence Effects in Double-Helical DNA III Average Properties of Curved DNA.* Biopolymers, 1988. **27**: p. 585-603.

177. Goodsell, D.S., et al., *Base pair roll and tilt in B-DNA bending*, in *Structural Biology: The State of the Art. Proceedings of the 8th Conversation.*, R.H. Sarma and R.H. Sarma, Editors. 1994, Adenine Press: Albany, New York. p. 215-220.

178. Zubay, G. and P. Doty, *Nucleic acid interactions with metal ions and amino acids.* Biochim Biophys Acta, 1958. **29**(1): p. 47-58.

179. Warrant, R.W. and S.H. Kim, *alpha-Helix-double helix interaction shown in the structure of a protamine-transfer RNA complex and a nucleoprotamine model.* Nature, 1978. **271**(5641): p. 130-5.

180. Warrant, R.W. and S.-H. Kim, Nature, 1978. **271**: p. 130.

181. Mirzabekov, A.D. and A. Rich, *Asymmetric lateral distribution of unshielded phosphate groups in nucleosomal DNA and its role in DNA bending.* Proc Natl Acad Sci U S A, 1979. **76**(3): p. 1118-21.

182. Carter, C.W., Jr. and J. Kraut, *A proposed model for interaction of polypeptides with RNA.* Proc Natl Acad Sci U S A, 1974. **71**(2): p. 283-7.

183. Seeman, N.C., J.M. Rosenberg, and A. Rich, *Sequence-specific recognition of double helical nucleic acids by proteins.* Proc. Natl. Acad. Sci. U. S. A, 1976. **73**(3): p. 804-808.

184. Berman, H.M., et al., *The Protein Data Bank.* Nucleic Acids Research, 2000. **28**: p. 235-242.

185. Dickerson, R.E. and T.K. Chiu, *Helix bending as a factor in protein/DNA recognition.* Biopolymers, 1997. **44**(4): p. 361-403.

186. Dickerson, R.E., *DNA bending: the prevalence of kinkiness and the virtues of normality.* Nucleic Acids Res, 1998. **26**(8): p. 1906-26.

187. Nadassy, K., S.J. Wodak, and J. Janin, *Structural features of protein-nucleic acid recognition sites.* Biochemistry, 1999. **38**(7): p. 1999-2017.

188. Jones, S., et al., *Protein-DNA interactions: A structural analysis.* Journal of Molecular Biology, 1999. **287**(5): p. 877-896.

189. Zakrzewska, K., R. Lavery, and B. Pullman, *Theoretical studies on the interaction of proteins and nucleic acid. II. The binding of alpha-helix to B-DNA.* Biophys. Chem, 1986. **25**(2): p. 201-13.

190. DiCapua, F., *Molecular Dynamics and Monte Carlo Studies of Protein Stability and Protein-DNA Interactions.* 1991, Wesleyan University.

191. Jen-Jacobson, L., L.E. Engler, and L.A. Jacobson, *Structural and thermodynamic strategies for site-specific DNA binding proteins.* Structure, 2000. **8**(10): p. 1015-23.

192. Jen-Jacobson, L., *Structural-perturbation approaches to thermodynamics of site-specific protein-DNA interactions.* Methods in Enzymology, 1995. **259**: p. 305-44.

193. Jen-Jacobson, L., *Protein-DNA recognition complexes: conservation of structure and binding energy in the transition state.* Biopolymers, 1997. **44**(2): p. 153-80.

194. Jayaram, B., et al., *Free Energy Analysis of Protein-DNA Binding: The EcoRI Endonuclease - DNA Complex.* J. Comput. Phys., 1999. **151**(1): p. 333-357.

195. Jayaram, B., et al., *Free-energy component analysis of 40 protein-DNA complexes: A consensus view on the thermodynamics of binding at the molecular level.* Journal of Computational Chemistry, 2002. **23**(1): p. 1-14.

196. Tang, Y. and L. Nilsson, *Molecular dynamics simulations of the complex between human U1A protein and hairpin II of U1 small nuclear RNA and of free RNA in solution.* Biophys J, 1999. **77**(3): p. 1284-305.

197. Sen, S. and L. Nilsson, *Free energy calculations and molecular dynamics simulations of wild- type and variants of the DNA-EcoRI complex.* Biophys J, 1999. **77**(4): p. 1801-10.

198. Sen, S. and L. Nilsson, *Structure, interaction, dynamics and solvent effects on the DNA-EcoRI complex in aqueous solution from molecular dynamics simulation.* Biophys J, 1999. **77**(4): p. 1782-800.

199. Zakrzewska, K. and R. Lavery, *Modelling Protein-DNA Interactions.* Theoretical Computational Chemistry, 1999. **8**: p. 441-483.

200. Matthews, B.W., *Protein-DNA interaction. No code for recognition [news].* Nature, 1988. **335**(6188): p. 294-295.

201. Benos, P.V., A.S. Lapedes, and G.D. Stormo, *Is there a code for protein-DNA recognition? Probab(ilistical)ly.* Bioessays, 2002. **24**(5): p. 466-75.

202. Williamson, J.R., *Induced fit in RNA-protein recognition.* Nat Struct Biol, 2000. **7**(10): p. 834-7.

203. Record, M.T., Jr., W. Zhang, and C.F. Anderson, *Analysis of effects of salts and uncharged solutes on protein and nucleic acid equilibria and processes: a practical guide to recognizing and interpreting polyelectrolyte effects, Hofmeister effects, and osmotic effects of salts.* Adv Protein Chem, 1998. **51**: p. 281-353.

204. Anderson, C.F. and M.T. Record, Jr., *Salt-nucleic acid interactions.* Annu Rev Phys Chem, 1995. **46**: p. 657-700.

205. Kombo, D.C., et al., *Computational analysis of variants of the operator binding domain of the bacteriophage lambda repressor.* International Journal of Quantum Chemistry, 1999. **75**(3): p. 313-325.

206. Kombo, D.C., M.A. Young, and D.L. Beveridge, *One Nanosecond Molecular Dynamics Simulation of the N-Terminal Domain of the λ–Repressor Protein.* Biopolymers, 2000. **53**: p. 596-605.

207. Kombo, D.C., M.A. Young, and D.L. Beveridge, *Molecular dynamics simulation accurately predicts the experimentally-observed distributions of the (C, N, O) protein atoms around water molecules and sodium ions.* Proteins, 2000. **39**(3): p. 212-215.

208. Kombo, D.C., M.A. Young, and D.L. Beveridge, *One nanosecond molecular dynamics simulation of the N-terminal domain of the lambda repressor protein.* Biopolymers, 2000. **53**(7): p. 596-605.

209. Kombo, D.C., et al., *Molecular dynamics simulation reveals sequence-intrinsic and protein-induced geometrical features of the OL1 DNA operator.* Biopolymers, 2001. **59**(4): p. 205-25.

210. Kombo, D.C., et al., *Calculation of the Affinity of the lamda Repressor-Operator Complex Based on free Energy Component Analysis.* Molecular Simulation, 2002. **28**: p. 187-211.

211. Dixit, S.B., D.Q. Andrews, and D.L. Beveridge, *Conformational Entropy In Protein-DNA Complexes: Estiamtes from Molecular Dynamics Simulations.* MS in Prep., 2003.

212. Pitici, F., D.L. Beveridge, and A.M. Baranger, *Molecular dynamics simulation studies of induced fit and conformational capture in U1A-RNA binding: Do molecular substates code for specificity?* Biopolymers, 2002. **65**(6): p. 424-35.

213. Kapanidis, A.N., et al., *Mean DNA bend angle and distribution of DNA bend angles in the CAP-DNA complex in solution.* J Mol Biol, 2001. **312**(3): p. 453-68.

214. Tchernaenko, V., H.R. Halvorson, and L.C. Lutter, *Topological measurement of an A-tract bend angle: variation of duplex winding.* J Mol Biol, 2003. **326**(3): p. 751-60.

215. Tchernaenko, V., et al., *Topological measurement of an A-tract bend angle: comparison of the bent and straightened states.* J Mol Biol, 2003. **326**(3): p. 737-49.

216. Rozenberg, H., et al., *Structural code for DNA recognition revealed in crystal structures of papillomavirus E2-DNA targets.* Proc Natl Acad Sci U S A, 1998. **95**(26): p. 15194-9.

217. Hegde, R.S., *The papillomavirus E2 proteins: structure, function, and biology.* Annu Rev Biophys Biomol Struct, 2002. **31**: p. 343-60.

218. Perez-Canadillas, J.M. and G. Varani, *Recent advances in RNA-protein recognition.* Curr Opin Struct Biol, 2001. **11**(1): p. 53-8.

219. Mount, D.W., *Bioinformatics: Sequence and Genome Analysis*. 2001, Cold Spring Harbor, NY: Cold Spring Harbor Laboratory Press.

220. Lafontaine, I. and R. Lavery, *ADAPT: A molecular mechanics approach for studying the structural properties of long DNA sequences.* Biopolymers, 2000. **56**(4): p. 292-310.

221. Lafontaine, I. and R. Lavery, *Optimization of nucleic acid sequences.* Biophys J, 2000. **79**(2): p. 680-5.

222. Lafontaine, I. and R. Lavery, *High-speed Molecular Mechanics Searches for Optimal DNA Interaction Sites.* Comb Chem High Throughput Screen, 2001. **4**(8): p. 707-17.

223. Thayer, K.M. and D.L. Beveridge, *Hidden Markov models from molecular dynamics simulations on DNA.* Proc Natl Acad Sci U S A, 2002. **99**(13): p. 8642-7.

224. Gaston, K., A. Kolb, and S. Busby, *Binding of the Escherichia coli cyclic AMP receptor protein to DNA fragments containing consensus nucleotide sequences.* Biochem J, 1989. **261**(2): p. 649-53.

225. Prevost, C., et al., *Persistence analysis of the static and dynamical helix deformations of DNA oligonucleotides: application to the crystal structure and molecular dynamics simulation of d(CGCGAATTCGCG)2.* Biopolymers, 1993. **33**(3): p. 335-350.

226. Stormo, G.D. and D.S. Fields, *Specificity, free energy and information content in protein-DNA interactions.* Trends Biochem Sci, 1998. **23**(3): p. 109-13.

227. Busby, S. and R.H. Ebright, *Transcription activation by catabolite activator protein (CAP).* J Mol Biol, 1999. **293**(2): p. 199-213.

228. Schlitter, J., *Estimation of Absolute and Relative Entropies of Macromolecules Using the Covariance Matrix.* Chem. Phys. Lett., 1993. **215**: p. 617-621.

Chapter 3

The Structural Basis for A-Tract DNA Helix Bending

Douglas MacDonald[1] and Ponzy Lu[2]

[1]Biolgie Structurale et Chimie, L'Institut Pasteur, Paris, France
[2]Department of Chemistry, University of Pennsylvania,
Philadelphia, PA 19104

Recent solution structures of A-tract DNA bends determined
by Nuclear Magnetic Resonance Spectroscopy using residual
dipolar couplings have provided new structural information
about the molecular basis of A-tract induced DNA bending.
We present a model of A-tract curvature that is consistent with
last twenty years of biophysical experiments performed on A-
tract DNA.

Introduction

Unusually slow migration of phased tandem A-tracts in gel electrophoresis
as compared to DNA fragments of equal length has been a puzzle for more than
twenty years. This debate has been due to the lack of a unique structure of an
A-tract DNA bend consistent with solution experiments. Single crystal x-ray
diffraction studies have been tainted by the tyranny of the lattice[1,2]. These
structures show that A-tract curvature, in the crystal, adopts several
conformations for identical sequences. Their direction of curvature is the result
of their orientation in the crystal. Nuclear magnetic resonance spectroscopy

(NMR) structures, in the past, have been calculated using only distance restraints from Nuclear Overhauser effects and torsion angle restraints obtained from scalar couplings[3]. Because this information is limited to about 5 Å, propagation of errors in these local restraints along the helix axis makes definition of a long-range helix path difficult. Recently however methods that result in anisotropic environments for nucleic acids in solution have resulted in the measurement of residual dipolar couplings (RDCs) of these biomolecules in the NMR spectrometer[4-6]. These RDCs yield the angular orientation of numerous bond vectors relative to a universal alignment tensor. When combined with distance and torsion angle restraints long-range characteristics can be defined. Recently RDCs have been used to determine several structures of DNA helices containing A-tracts[7,8]. These new solution structures along with the four previous x-ray structures[1,2,9] allow us to describe a structural basis for A-tract induced helix bending.

Characteristics of A-tracts in Solution

The question we wish to answer in the current work is how homopolymeric (dA·dT) runs (A-tracts) induce curvature along a DNA double helix. The structure must be able to explain all of the past biophysical solution experiments performed with A-tract DNA and its tandem concatamers. The most salient characteristics of A-tracts in solution are: (1) A preferential direction of bending within A-tract sequences[10]. Maximum curvature is only produced when A-tracts are repeated at a constant phasing corresponding to the helical repeat of DNA (~10-10.5 basepairs/turn). Phasing slightly less or greater then the helical repeat produces concatamers with less overall curvature. Additionally, non-vectorial flexibility within phased A-tracts can be ruled out as a source of curvature since sequences containing five AT base-pairs, A_5, ligated in helical repeat, $(A_5N_5)_n$, produce substantial curvature. While A_5 tracts ligated at 1.5 helical repeats, $(A_5N_{10})_n$ produce no detectable curvature. (2) Ligated A-tracts in solution have an imperfect 2-fold axis of bending that curves toward the minor groove of A-tracts or in the direction of tilt at the junctions[10]. Rotation of every other A-tract by 180° to yield $(A_6N_4T_6N_4)_n$ multimers produces no substantial difference in gel mobility (see figure 1a). This result rules out base-pair roll at the junction between A-tracts and the adjoining DNA since the resulting geometry of curvature would not produce a 2-fold axis of curvature, and hence ligated $A_6N_4T_6N_4$ tracts would not constructively add to form a curved super-helical structure (see figure 1b & 1c). However models with positive roll distributed over adjacent basepairs within the GC rich region between A-tracts are consistent with this imperfect 2-fold axis of curvature. (3) The 5' and 3' junctions between A-tracts and the surrounding DNA are not equivalent. Multimers of the form ($A_4T_4N_2$) are highly curved, whereas those of the form ($T_4A_4N_2$) are straight[11]. In addition hydroxyl radical cleavage patterns suggest a progressively narrowing of the minor groove from the 5' to 3' end within A-tracts in solution[12]. (4) Disruption of the run of A's within an A-tract by a G,C, or T base produces a gel migration pattern, of the respective substituted concatamers, that has near normal gel mobility[10] (see figure 1a). This implies that A-tracts function as a cooperative unit in which a continuous run of A's is necessary to observe strong bending anomalies. (5) The electrophoretic mobility of phased A-tracts is not drastically affected by the sequence within the GC containing regions connecting adjacent A-tracts[13].

Models of A-tract Curvature

These five properties of A-tracts, without additional structural information, suggest three models of curvature as possible candidates to explain A-tract induced curvature. The first model proposes that the A-tract region exists in an alternative non B-DNA form with base pairs at a negative inclination relative to the overall helix axis[10]. When this unusual DNA structure meets the adjoining B-DNA at the junctions, curvature in the helical axis is produced (see figure 2a). This junction model is similar to those proposed for the interface of A- and B-DNA[14]. The second model, or wedge formalism, suggests that helix axis curvature is produced by uniform negative roll angles at ApA steps within the A-tract[15,16]. Here, unlike the junction model, the axis of curvature is defined at the level of dinucleotide steps. Curvature is produced when this curved poly (dA·dT) segment is flanked, in the double helix, with straight B-DNA (see figure 2b). Although this model assumes uniform negative roll angles between ApA steps there still must be some form of cooperativity between neighboring bases within the A-tract to account for the non linear increase in curvature when the adenine repeat is increased from 3 to 4 A's in tandem[10]. The former shows almost no curvature while the latter has substantial curvature in solution. The third model is the general sequence model of Dickerson and colleagues[17-22]. This model differs from the junction and wedge models in that curvature is not caused by A-tracts but stems from the constructive addition of the natural writhe of B DNA. A-tracts, that are straight and B-form, simply provided the correct phasing for this constructive addition to occur (see figure 2c).

The Direction of Curvature

Alone no one of these models can be preferentially selected over another without additional structural information. In the past this meant that all 3

(a)

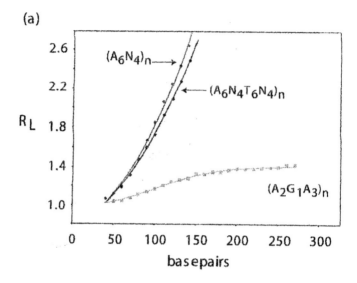

Figure 1. Geometry of Curvature. (a) Plot of the R_L (relative gel electrophoretic mobility) values vs. actual chain length for three different DNA sequences. Notice that a 180° rotation of every other A-tract to produce $(A_6N_4T_6N_4)_n$ does not change the mobility of the ligated concatamers relative to $(A_6N_4)_n$. Additionally interruption of an A-tract by a G (shown), C or T base produces a helix with near normal mobility. (b) and (c) are schematic diagrams showing that only when curvature occurs in the direction of tilt at the junctions is the axis of curvature 2-fold symmetric, i.e. rotation of an A-tract by 90° about its helical axis reveals that curvature lies only in one plane. Hence A tracts and T tracts that are adjacent to each other will produce substantial curvature. Curvature in the direction of roll at the junctions, like all x-ray structures of A-tracts, is not 2-fold symmetric.

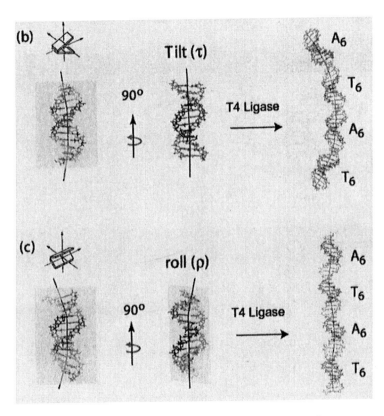

Figure 1. *Continued. (See page 1 of color insert.)*

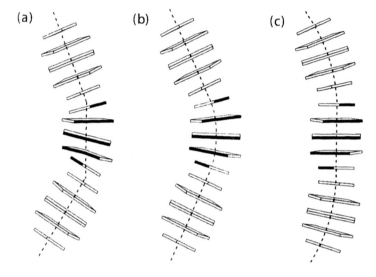

Figure 2. Three models of A-tract curvature. Each basepair within a helix is represented by rectangular blocks. AT basepairs have rectangular blocks with black edges. The junction model, (a), proposes that the AT bases adopt a non B-DNA structure. When this non B-DNA adjoins normal DNA curvature is produced at the junctions. The wedge formalism, (b), suggests that curvature is produced by uniform negative roll angles at ApA steps within the A-tract. The general sequence model, (c), asserts that curvature is produced outside the A-tract region due to the constructive addition of the natural writhe of DNA.

models were valid candidates to explain A-tract induced curvature. Recently, however, NMR structures[7,8] determined on A-tract DNA using residual dipolar couplings provides the additional structural information necessary to choose one of the above models (or a composite) as the mechanism of A-tract induced curvature in solution. One of the new NMR structures, A_6, along with the four previously determined X-ray structures are shown in figure 3. The NMR structure, (A_6)[8], is shown at the left while the four previous x-ray structures of A-tracts are present at the right for comparison: the Nelson[9], (N), the up helix[1], (D_u), the down helix[1], (D_d), and the DiGabriele & Steitz[2], (D_a), structure. The third adenosine nucleotide of the A-tract in each helix is shown in red for orientation. In row A the deoxyribose of the third adenosine within the A-tract is viewed in line with the best helical axis as determined by CURVES 5.3[23,24]. Row B is a 270° rotation of the row A view about the z-axis. Comparing the NMR structure, A_6, with the four x-ray structures in row A clearly shows that the A_6 NMR structure has a 2-fold axis of bending that curves toward the minor groove of the A-tract consistent with previous biochemical experiments[25]. By contrast the Nelson, (N), and Uphelix, (D_u), show no overall bending in this view while the Downhelix, (D_d), and DiGabriele & Steitz, (D_a), structures show bends which are 180° away from that predicted by solution measurements. When viewing the five helices from row B, the NMR structure, A_6, shows no overall bend since it is in a plane perpendicular to the page. The four x-ray structures show overall bends that are in the plane of the page. To help visualize the difference in the direction of curvature between the NMR structure, A_6, and the four X-ray structures we have included normal vector diagrams[26] of these DNA fragments (see figure 4). These diagrams are generated by drawing a vector perpendicular to the best least squares plane of each basepair. These vectors are placed on a common origin and viewed along the helical axis. The inner and outer circles indicate 5° and 10° deviation from the reference axis, respectively. The arrows (3A) and (3B) indicate the eye placement in the views presented in Figure 3, rows A and B. The NMR structure, (A_6), when viewed from arrow 3A shows a sequential progress of graphed vectors, indicating an overall bend towards the minor groove of the A-tract. In contrast all of the four crystal structures show a progression of vectors that are 90° rotated from the bend in the NMR, A_6, structure. The direction of curvature in all four x-ray structures is not towards the minor groove of the A-tract. These structures when concatenated as alternating A and T tracts will not produce a curved super-helical structure (see figure 1).

The Mechanism of Induced Curvature

Although the x-ray structures are not useful in determining the direction of curvature produced by A-tracts in solution, because their overall direction of curvature depends on the sequence's orientation in the crystal, DiGabriele & Steitz pointed out that after superposition of the four A-tract regions a root mean square (rms) dispersion of only 0.46-0.69 Å for each atom is found[2]. This compares to a rms dispersion for all atoms of 1.51-2.67 Å (see table 1). These

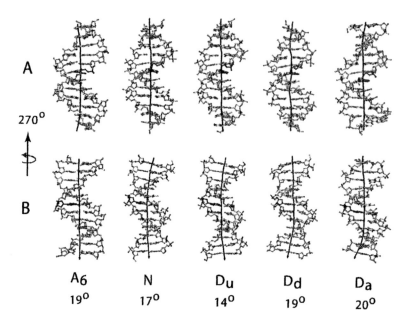

Figure 3. Comparison of A-tract DNA structures. The NMR structure, (A₆), is shown at the left while the four x-ray structures of A-tracts are present at the right for comparison: the Nelson, (N), the up helix, (Dᵤ), the down helix, (Dₐ), and the DiGabriele & Steitz, (Dₐ), structure. The third adenosine nucleotide of the A-tract in each helix is shown in red for orientation. In row A the deoxyribose of the third adenosine within the A-tract is viewed in line with the best helical axis as determined by CURVES 5.3. Row B is a 270° rotation of the row A view about the z-axis. Comparing the NMR structure, A₆, with the four x-ray structures in row A clearly shows that the A₆ NMR structure has a 2-fold axis of bending that curves toward the minor groove of the A-tract consistent with previous phasing experiments. By contrast the Nelson, (N), and Uphelix, (Dᵤ), show no overall bending in this view while the Downhelix, (Dₐ), and DiGabriele & Steitz, (Dₐ), structures show bends which are 180° away from that predicted by solution measurements. When viewing the five helices from row B, the NMR structure, A₆, shows no overall bend since it is in a plane perpendicular to the page. The four x-ray structures show overall bends that are in the plane of the page. (See page 2 of color insert.)

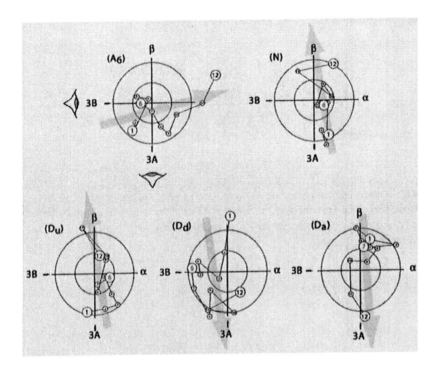

Figure 4. Normal-vector plots of A-tract DNA Structures. These diagrams are generated by drawing a vector perpendicular to the best least squares plane of each base pair. These vectors are placed on a common origin and viewed along the helical axis. The inner and outer circles indicate 5° and 10° deviation from the reference axis, respectively. The arrows (3A) and (3B) indicate the eye placement in the views presented in Figure 3, rows A and B. The NMR structure, (A_6), when viewed from arrow 3A shows a sequential progress of graphed vectors indicating an overall bend towards the minor groove of the A-tract. In contrast all of the four crystal structures show a progression of vectors that are 90° rotated from the bend in the NMR, A_6, structure. The direction of curvature in all four x-ray structures is not towards the minor groove of the A-tract. (See page 2 of color insert.)

Table 1. Comparison A-tract Stuctures by superposition rms distances ($\overset{o}{A}$)

	A6 NMR	Nelson	Up Helix	Down Helix	DiGab. Steitz
A6 NMR	------	**2.70**	**2.56**	**2.62**	**2.64**
Nelson	3.86	------	**0.52**	**0.69**	**0.54**
Uphelix	4.00	1.51	------	**0.67**	**0.46**
Downhelix	4.14	2.42	1.92	------	**0.60**
DiGab./Steitz	3.67	2.45	2.67	2.30	------

Distances in normal and bold text give rms ($\overset{o}{A}$) differences between the complete structures or the A-tract region only, respectively.

larger differences result from a fundamental difference in geometry between the solution and crystal structure A-tracts. Figure 5a shows a plot of the base inclination for the A-strands of the A_6 NMR[8] (red), the Nelson[9] (green), the up helix[1] (blue), the down helix[1] (blue dashed), and DiGabriele & Steitz[2] (pink). Figure 5b compares the inclination of the A_6 NMR (red) and the $A_2G_1A_3$ NMR[8] (blue). The adenine bases within the A_6 NMR structure are negatively inclined relative to the global axis while the adenine bases from the crystal structures are not. This negative inclination initiates in the middle of the row of A's and increases as one moves in the 5' to 3' direction. This increase in inclination does not result in tilt. Instead the corresponding thymine bases continue to stack with their neighbors resulting in a large buckle at the 3' end of the A-tract (see figure 5c). Why does the geometry of the A-tracts found in the crystal structures differ from that of the solution structure? In the past crystal packing forces as well as 2-methyl-2-4-pentanediol (MPD) concentration were suggested to be the source of perturbation that caused the x-ray structures geometry to be different then that predicted in solution[22,27,28]. This could also be the reason for the difference in geometry between the A_6 NMR structure and that of the four X-ray structures. But a recent crystal structure of the DNA dodecamer (dACCGAATTCGGT) shows a base stacking geometry that is similar to the A_6 NMR structure[29]. This structure's central abbreviated A-tract region, AATT, has bases that are negatively inclined relative to the helix axis. This would suggest that at least in short A-tract sequences the crystal lattice and/or MPD concentration does not interfere with the geometry of A-tract DNA.

How does this change in inclination cause A-tract induced curvature? Figure 6a & 6b show the accessible surface areas, viewed from the center of each A-tract's minor groove, for both the A_6 and $A_2G_1A_3$ NMR structures. The two sets of three basepairs in stick representation at the bottom of the panels correspond to the 3' end of each structure, minus the terminal basepairs. Adenine bases are shown in red, guanines in blue, cytosines in yellow and thymine bases in aqua. The two bases at position six, an adenine for the A_6 structure and a guanine for the $A_2G_1A_3$ NMR structure are depicted to remind the reader that only the basepairs at this position are different between the two structures. Any change in geometry between the two structures must be a result of this AT to GC transition at position 6. Upon inspection of the 3' end of both structures it becomes clear that the uninterrupted A-tract has a larger change in base inclination between base C10 and G11 at its 3' junction then the corresponding $A_2G_1A_3$ NMR structure (also see figure 5b). This larger change in inclination directly translates into a large roll within its GC rich region, C10-G11 step 6.0° & G11-G12 step 7.2°, compared to the $A_2G_1A_3$ NMR structure, C10-G11 step 3.4° & G11-G12 step 3.4° (see panels 6c & d). The reason for the roll in both structures is that the change in inclination at the junction results in an energetically unstable base stacking arrangement. To alleviate this, the bases within the non A-tract region roll into the major groove. This explains why the base sequence between A-tracts has only a slight effect, ±10%, on overall curvature[13]. The roll occurs in response to the change in inclination not base sequence.

Why do A-tracts require a minimum number of adenine bases to produce curvature? Koo *et al.* noticed in 1986 that at least four adenine bases in tandem where required to obtain significant curvature with six adenine bases yielding the largest detected curvature[10]. Clearly the difference in curvature between these short and long A-tracts must originate from some structural difference. Figure 7 shows the distances within the minor groove for A-tracts containing different numbers of adenine bases. The notation is that of Nadeau and Crothers[30]. Inspection of the type a distances, an indication of minor groove width, shows that the minor groove of A-tracts become increasingly narrow as the length of the adenine repeat is increased, with the minor groove width remaining constant when six or more adenines in tandem are present. This finding was predicted by early NMR measurements and reflects the increase in propeller twist as well as inclination within adenine bases for longer A-tracts[30]. The second distance, labeled d, is an indication of base inclination. Here we see two distinct groups of distances. The first group, are short A-tracts, A_4 and $A_2G_1A_3$, which have a type d distance of ~ 4.5 Å. These sequences produce only limited curvature, 8-9°, because they are not long enough to fully stabilize the cumulative build up of negative inclination within their adenine bases. As figure 5b shows, the negative inclination within the A strand of the A_6 NMR structure does not become significant until the fifth adenine. Structures with the sequence, A_4, would not be expected to posses negative inclination within their limited adenine run. The recent NMR structure of an A_4 tract confirms this[7]. The same reasoning can explain the reduced curvature in the $A_2G_1A_3$ sequence. The interruption of the A-tract by the guanine base interferes with the build up of negative inclination as discussed in the previous section. The second group of type d distances is for A-tracts with five or more adenines in tandem. These

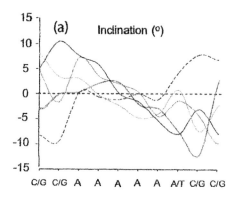

Figure 5. Base inclination within A-tracts. Graph, (a), shows the base inclination for the A-strands of the A_6 NMR (red), the Nelson (green), the Uphelix (blue), the Downhelix (blue dashed), and DiGabriele & Steitz (pink). Only the adenine bases within the A_6 NMR structure show increasing negative inclination as one moves from the center of the A-tract to the 3' end. (b) Shows the base inclination for the A-strands of the A_6 NMR (red) and the $A_2G_1A_3$ NMR (blue). Notice how the AT to GC transition in the middle of the A-tract eliminates the negative inclination at the 3' end of the A-tract. Graph, (c), plots buckle for the A_6 NMR (red) and the $A_2G_1A_3$ NMR (blue). The large inclination values within the A-strand for the A_6 NMR (red) structure do not produce tilt. Instead the corresponding thymine bases continue to stack with their neighbors resulting in a large positive buckle at the 3' end of the A-tract (See page 3 of color insert.)

Continued on next page.

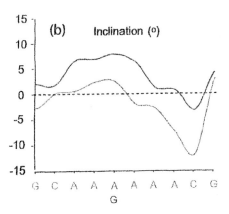

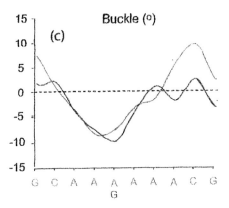

Figure 5. *Continued* *(See page 3 of color insert.)*

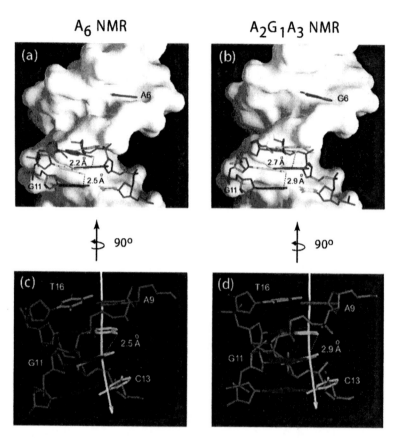

A₆ NMR   A₂G₁A₃ NMR

Figure 6. (a) and (b) The accessible surface areas of the A₆ and A₂G₁A₃ NMR structures, viewed from the center of each A-tract's minor groove. The two sets of three basepairs in stick representation at the bottom of the panels correspond to the 3' end of each structure, minus the terminal basepairs. Adenine bases are shown in red, guanines in blue, cytosines in yellow and thymine bases in aqua. The two bases at position six, an adenine for the A₆ structure and a guanine for the A₂G₁A₃ NMR structure are depicted to remind the reader that only the basepairs at this position are different between the two structures. Notice the greater change in inclination between C10 and G11 in the A₆ NMR structure relative to the A₂G₁A₃ structure. In (c) and (d), H8 of G11 points out roughly perpendicular to the plane of the paper. The hydrogen bond between G11 and C10 stabilizes the roll between the two bases. The larger roll and thus shorter hydrogen bond in the A₆ NMR structure is the direct result of the larger change in inclination in this structure. (See page 4 of color insert.)

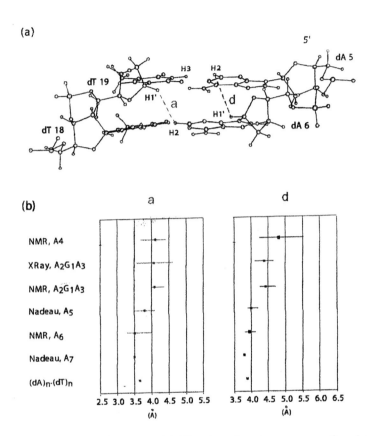

Figure 7. Minor groove distances. (a) View into the minor groove of an A-tract with the notation of Nadeau and Crothers[30]. (b) The two graphs show the average (square) and range (lines) for the distances in the various A-tracts. For Nadeau, A$_7$ only internal NOE distances were measured and thus no range is given. The fiber diffraction distances for (dA)$_n$·(dT)$_n$ are based on the model of Lipanov & Churpina calculated by Nadeau & Crothers. For the A$_2$G$_1$A$_3$ x-ray structure protons were added using Insight II (Accelrys Inc.)
(See page 5 of color insert.)

structures produce greater curvature because they accumulate the negative inclination within their adenine bases. This greater amount of inclination is reflected by the shorter type d distance of ~ 4 Å for these sequences.

Reasons for A-tract Curvature

Now that we have a structural basis for A-tract induced curvature we explain why A-tracts, unlike other sequences, are able to stabilize the negative inclination found within their adenine bases. We suggest three possible mechanisms that could act as this stabilizing force. (1) The formation of bifurcated hydrogen bonds[9,31], (2) a spine of hydration and/or monovalent cations within the minor groove[32,33] or (3) the stacking of the adenine bases so as to maximize their π overlap[34].

Bifurcated hydrogen bonds are not a significant force in the formation and/or stabilization of the negative inclination found within A-tracts, since their existence is not necessary for curvature. A-tracts that have one or two adenine bases replaced by the base analog inosine still show significant curvature even though these substituted sequences cannot form bifurcated hydrogen bonds[35,36]. These sequences instead have close inter-strand contacts at the major groove between amino groups across A-I basepair steps. In addition, sequences that contain as little as three adenines in tandem form bifurcated H-bonds although they exhibit no significant curvature in solution[31].

An ordered spine of hydration has been suggested for stabilization of B' DNA in the minor groove of A-tracts in solution[2,32]. The role of hydration on DNA structure is well documented, but does the hydration found within A-tracts differ from regular B DNA such that it could stabilize the B' form? Although determination of exact water coordination in solution is not yet possible[37] x-ray structures of A-tracts and sequences containing AT basepairs allows us to suggest coordination geometries for water molecules within the minor groove of A-tracts in solution. The first shell of water (dark blue in figure 8a & c) most probably coordinates between the N3 of the adenine and the O2 of thymine across the minor groove. The second shell (light blue in figure 8a & c) bridging the first to ensure that the water maintains its tetrahedral coordination. This coordination geometry would be the same as originally seen in the Dickerson Dodecamer and the asymmetric A-tract x-ray structure[2,38]. If this spine of hydration is responsible for the formation of B' DNA, then by breaking this spine we should be able to abolish curvature. Figure 8b & d shows the possible disruption of the spine of hydration within the minor groove of the $A_2G_1A_3$ NMR structure. The sequence of this structure is identical to the A_6 NMR structure except (see figure 8a & c) for an AT to GC transition at position six. This interruption reduces curvature from 19° in the uninterrupted A_6 NMR tract to 9° in the $A_2G_1A_3$ NMR tract[8]. This reduction in curvature corresponds with the results of cyclization experiments[10,39]. However if this spine of hydration is the main driving/stabilizing factor in B' DNA then poly inosine-cytosine tracts should produce curvature when ligated in phase since their atoms in the minor groove are isosteric with A-tract DNA[35]. Since, inosine-cytosine tracts do not

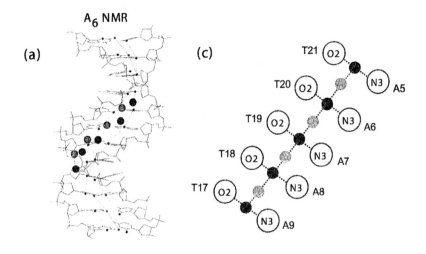

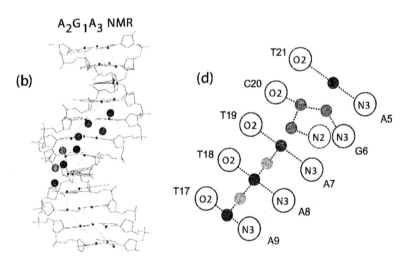

Figure 8. Minor Groove Hydration within A-tracts. (a) and (b) are DNA helices in stick representation for an A_6 and $A_2G_1A_3$ tract. (c) and (d) are corresponding schematic diagrams. First shell solvent is in dark blue, second shell in light blue and oxygens that disrupt the normal Dickerson Dodecamer pattern are in pink. Red and green circles represent atoms that are H-bond acceptors and donors respectively. Notice how the insertion of guanine base within an A-tract breaks the normal spine of hydration. (See page 6 of color insert.)

produce macroscopic curvature there must be an additional factor responsible for A-tract curvature.

Recently, the role of monovalent cations in DNA curvature has received renewed attention with the determination of several extremely high resolution x-ray structures[33,40-41]. In particular Williams has postulated that monovalent cations in the minor groove of A-tracts may be the source of DNA curvature in these sequences[33]. These cations would neutralize the phosphates causing the helix backbone to slightly collapse producing curvature towards the minor groove, a direction that is in agreement with solution studies[25]. In addition if cations also bound to the major groove, half a helical turn away, this would produce additional curvature in the proper direction. Thus this model of electrostatic collapse can readily explain the magnitude and direction of curvature of A-tract DNA. The question to be addressed is do these cations have a long enough residence time in the grooves of A-tract DNA to produce the curvature within A-tract DNA. Current solution methods measuring cation lifetimes in the grooves of DNA do not allow an unambiguous answer to this question. Solution experiments performed on A-tracts with respect to the electrostatic collapse (EC) model allows us to suggest that this mechanism of curvature is of secondary importance in A-tract DNA.

First, AT to TA transversions within the middle of an A-tract drastically reduce curvature within the helix[10]. This reduction in curvature according to the EC model must stem from a shortening of the residence time for the cations within the minor groove of A-tracts. A change in cation binding in the major groove one-half a helical turn away from a AT to TA transversion is highly unlikely. Since only the arrangement of the cations within the minor groove changes, the A-tract itself and not the flanking sequence must be largest contributor to curvature. This result is in contrast to all x-ray and NMR structural studies that show that A-tracts are relatively straight[1,2,8,9,42].

Second, free solution electrophoresis experiments show that the A_3T_3 motif has greater cation binding then an A_5 motif[43]. Thus using the electrostatic collapse model would suggest that a DNA helix containing A_3T_3 motifs repeated in helical phase would have a larger degree of curvature then the corresponding A_5 motif. The A_5 motif is more highly curved in solution.

The last factor to consider is purine-purine stacking within the A-tract. We believe this is the dominating factor driving the formation of B' DNA. As seen in all structures of full A-tracts the adenine bases maximize their π overlap producing basepairs with high propeller twist[1,2,8,9]. In contrast poly GC tracts with their three Watson Crick basepairs are not able to propeller twist to the same extent as AT basepairs resulting in less base to base overlap[44] and hence no accumulation of negative inclination. In addition the increase in inclination found within the adenine bases of A-tracts is compensated by the paired thymine bases buckle to reduce the strain. The hydrogen bond characteristics are not distorted beyond their normal range. In this model the bifurcated H-bonds and the spine of minor groove hydration as well as possible cation interactions provide additional stabilization energy to the B' structure but by themselves are not sufficient to initiate the formation of negative inclination within A-tract DNA. The reasons for the lack of curvature seen in poly dI·C sequences could arise from differences in their major groove atoms. Although the stereochemistry of the major groove atoms does not effect curvature as drastically as their minor groove counterparts since the energy between normal B DNA and the B' DNA is not large, an enthalpy of 3.5 kcal/mol·AT pair[45], the lack of 5' methyl groups on the cytosine bases as well as the smaller oxygen group relative to an amine group at position 6 on the inosine base could make the formation of B' DNA less favorable[34,45].

A model of A-tract Curvature

As discussed in the above sections recent NMR structures as well as the existing four x-ray structures on A-tracts has allowed us to propose a mechanism of bending for A-tract DNA. In this section we present an overall view of A-tract induced bending in solution keeping in mind that the solution and crystal structures only provided a static view of a dynamic process. We start from the perspective that the A-tract DNA forms an alternative structure called B' DNA[10]. This structure differs from normal B DNA in that its 3' adenine bases possess negative inclination. Our model is an extension of the junction model. However, rather then with tilt within the A-tract, the thymine bases remain stacked with their neighbors resulting in large buckle at the A-tract junction. Like the general sequence model [17-22] of A-tract curvature, our model agrees in that most of the curvature occurs through roll in the GC containing intervening sequence. Our model does not support the wedge formulism[15,16] as the primary mechanism of curvature.

How does the model apply to A-tracts of different length? First A_4 tracts, the shortest A-tracts that produce abnormal gel mobility when ligated in phase[10]. The model results in A_4 tracts as well as longer A-tracts that are relatively straight having only slight curvature towards the minor groove due to negative roll between AT basepairs. On average this curvature is 1° per base step, or ~ 6° for an A_6 tract. This relatively straight A-tract structure is in agreement with x-ray crystallography, NMR spectroscopy and molecular dynamics simulations[1,2,8,9,42]. Second due to the increased propeller twist in all A-tracts and the formation of bifurcated hydrogen bonds in longer A-tracts, runs of poly dAT are stiffer then normal B DNA[9]. This rigidity, especially in shorter A-tracts that lack fully stabilized negative base inclination, may allow some global curvature to occur through the constructive addition of the natural writhe of B

DNA as proposed by the general sequence model. As the A-tract lengthens from A_4 to A_5 and A_6 the solution structure is increasingly characterized by negative inclination within the adenine bases. The observed slow gel mobility of these sequences when ligated in phase may still arise from the slight negative roll within the A-tract, and/or the proper phasing of the natural writhe of the GC rich intervening sequences. However most of the curvature is due to the change in inclination at the 3' junction. This results in positive roll dispersed throughout the intervening flanking sequences to alleviate unfavorable base stacking which would be present at the junctions in the absence of this roll.

Considering the large body of biochemical experiments performed on A-tracts with respect to our model the following picture emerges. Overall A-tract induced curvature is towards the minor groove of the A-tract resulting from an imperfect 2-fold axis of bending. This direction of curvature is in agreement with solution studies[10,25]. This imperfect axis of curvature results because the majority of bending takes place at the 3' junction. This explains why substitution of 2,6 diaminopurine (DAP) in the 5' end of A-tracts has a much smaller effect on anomalous gel migration then substitution in the 3' end of the A-tracts[46]. The latter substitution destroys the negative inclination, thereby abolishing the larger bend at the 3' junction. In addition curvature is drastically reduced when an A-tract is disrupted by a G,C or T base[10]. This non-adenine base when inserted in the middle of the A-tract disrupts the negative inclination thus reducing curvature. In contrast the general sequence model would necessitate a helical deflection of 9° towards the major groove within the A-tract to account for the observed decrease in bending. Crystallography as well as NMR provides no evidence for such a deflection. The x-ray structure of the DNA analog of the primer for HIV-1 reverse transcriptase that contains the $A_2G_1A_3$ sequence and the $A_2G_1A_3$ NMR structure both posses A-tract regions that lack significant curvature[8,47].

To explain why the DNA between A-tracts does not drastically affect curvature we must remember that the bending within these regions only occurs to restore favorable base stacking geometry between the A-tract and this adjoining B-DNA, not because of base sequence. Thus increasing the length of the linker DNA segment from 0.5 to 1.5 helical turns does not change the amount of curvature per A-tract[48]. In this extended linker region curvature takes place only in the first few basepairs. After favorable base stacking is achieved the remaining DNA on average will possess no curvature.

In conclusion we have suggested a structural basis for A-tract curvature that is consistent with the last twenty years of biophysical experiments performed on A-tract DNA. This mechanism assumes that A-tract DNA on average exists in a B' form with negative inclination within its 3' adenine bases. When these negatively inclined bases meet normal B-DNA at the 3' junction positive roll occurs, within the B-DNA, to alleviate unfavorable base stacking that would be present in the absence of this roll. The resulting overall direction of curvature is towards the minor groove of the A-tract.

References

1. Digabriele, A. D.; Sanderson, M. R.; Steitz, T. A. *Proc. Natl. Acad. Sci. U. S. A.* **1989**, *86*, 1816-1820.
2. Digabriele, A. D.; Steitz, T. A. *J. Mol. Biol.* **1993**, *231*, 1024-1039.
3. Wurthrich, K. *NMR of Proteins and Nucleic Acids*; John Wiley & Sons: New York, 1986.
4. Tolman, J. R.; Flanagan, J. M.; Kennedy, M. A.; Prestegard, J. H. *Proc Natl Acad Sci U S A* **1995**, *92*, 9279-9283.
5. Tjandra, N.; Bax, A. *Science* **1997**, *278*, 1111-1114.
6. Hansen, M. R.; Mueller, L.; Pardi, A. *Nat. Struct. Biol.* **1998**, *5*, 1065-1074.
7. Barbic, A.; Zimmer, D. P.; Crothers, D. M. *Proc Natl Acad Sci U S A* **2003**, *100*, 2369-2373.
8. MacDonald, D.; Herbert, K.; Zhang, X. L.; Pologruto, T.; Lu, P. *J. Mol. Biol.* **2001**, *306*, 1081-1098.
9. Nelson, H. C. M.; Finch, J. T.; Luisi, B. F.; Klug, A. *Nature* **1987**, *330*, 221-226.
10. Koo, H. S.; Wu, H. M.; Crothers, D. M. *Nature* **1986**, *320*, 501-506.
11. Hagerman, P. J. *Nature* **1986**, *321*, 449-450.
12. Burkhoff, A. M.; Tullius, T. D. *Cell* **1987**, *48*, 935-943.
13. Haran, T. E.; Kahn, J. D.; Crothers, D. M. *J. Mol. Biol.* **1994**, *244*, 135-143.
14. Selsing, E.; Wells, R. D.; Alden, C. J.; Arnott, S. *J Biol Chem* **1979**, *254*, 5417-5422.
15. Ulanovsky, L. E.; Trifonov, E. N. *Nature* **1987**, *326*, 720-722.
16. Trifonov, E. N.; Sussman, J. L. *Proc. Natl. Acad. Sci. U. S. A.* **1980**, *77*, 3816-3820.
17. Calladine, C. R.; Drew, H. R.; McCall, M. J. *J. Mol. Biol.* **1988**, *201*, 127-137.
18. Maroun, R. C.; Olson, W. K. *Biopolymers* **1988**, *27*, 585-603.
19. Goodsell, D. S.; Kopka, M. L.; Cascio, D.; Dickerson, R. E. *Proc. Natl. Acad. Sci. U. S. A.* **1993**, *90*, 2930-2934.
20. Goodsell, D. S.; Kaczorgrzeskowiak, M.; Dickerson, R. E. *J. Mol. Biol.* **1994**, *239*, 79-96.
21. Grzeskowiak, K.; Goodsell, D. S.; Kaczorgrzeskowiak, M.; Cascio, D.; Dickerson, R. E. *Biochemistry* **1993**, *32*, 8923-8931.
22. Dickerson, R. E.; Goodsell, D. S.; Neidle, S. *Proc. Natl. Acad. Sci. U. S. A.* **1994**, *91*, 3579-3583.
23. Ravishanker, G.; Swaminathan, S.; Beveridge, D. L.; Lavery, R.; Sklenar, H. *J. Biomol. Struct. Dyn.* **1989**, *6*, 669-699.

24. Lavery, R.; Sklenar, H. *J. Biomol. Struct. Dyn.* **1988**, *6*, 63-91.
25. Zinkel, S. S.; Crothers, D. M. *Nature* **1987**, *328*, 178-181.
26. Dickerson, R. E.; Kopka, M. L.; Pjura, P. *Proc. Natl. Acad. Sci. U. S. A.* **1983**, *80*, 7099-7103.
27. Dickerson, R. E.; Goodsell, D.; Kopka, M. L. *J Mol Biol* **1996**, *256*, 108-125.
28. Ganunis, R. M.; Guo, H.; Tullius, T. D. *Biochemistry* **1996**, *35*, 13729-13732.
29. Hizver, J.; Rozenberg, H.; Frolow, F.; Rabinovich, D.; Shakked, Z. *Proc Natl Acad Sci U. S. A.* **2001**, *98*, 8490-8495.
30. Nadeau, J. G.; Crothers, D. M. *Proc. Natl. Acad. Sci. U. S. A.* **1989**, *86*, 2622-2626.
31. Coll, M.; Frederick, C. A.; Wang, A. H. J.; Rich, A. *Proc. Natl. Acad. Sci. U. S. A.* **1987**, *84*, 8385-8389.
32. Chuprina, V. P. *Nucleic Acids Res.* **1987**, *15*, 293-311.
33. Woods, K. K.; McFail-Isom, L.; Sines, C. C.; Howerton, S. B.; Stephens, R. K.; Williams, L. D. *J. Am. Chem. Soc.* **2000**, *122*, 1546-1547.
34. Diekmann, S.; Mazzarelli, J. M.; McLaughlin, L. W.; von Kitzing, E.; Travers, A. A. *J Mol Biol* **1992**, *225*, 729-738.
35. Koo, H. S.; Crothers, D. M. *Biochemistry* **1987**, *26*, 3745-3748.
36. Diekmann, S.; von Kitzing, E.; McLaughlin, L.; Ott, J.; Eckstein, F. *Proc Natl Acad Sci U. S. A.* **1987**, *84*, 8257-8261.
37. Liepinsh, E.; Otting, G.; Wuthrich, K. *Nucleic Acids Res* **1992**, *20*, 6549-6553.
38. Drew, H. R.; Dickerson, R. E. *J Mol Biol* **1981**, *151*, 535-556.
39. Koo, H. S.; Drak, J.; Rice, J. A.; Crothers, D. M. *Biochemistry* **1990**, *29*, 4227-4234.
40. Tereshko, V.; Minasov, G.; Egli, M. *J. Am. Chem. Soc.* **1999**, *121*, 470-471.
41. Tereshko, V.; Minasov, G.; Egli, M. *J. Am. Chem. Soc.* **1999**, *121*, 3590-3595.
42. McConnell, K. J.; Beveridge, D. L. *J Mol Biol* **2001**, *314*, 23-40.
43. Stellwagen, N. C.; Magnusdottir, S.; Gelfi, C.; Righetti, P. G. *J. Mol. Biol.* **2001**, *305*, 1025-1033.
44. McCall, M.; Brown, T.; Kennard, O. *J Mol Biol* **1985**, *183*, 385-396.
45. Chan, S. S.; Breslauer, K. J.; Austin, R. H.; Hogan, M. E. *Biochemistry* **1993**, *32*, 11776-11784.
46. Mollegaard, N. E.; Bailly, C.; Waring, M. J.; Nielsen, P. E. *Nucleic Acids Res.* **1997**, *25*, 3497-3502.
47. Han, G. W.; Kopka, M. L.; Cascio, D.; Grzeskowiak, K.; Dickerson, R. E. *J Mol Biol* **1997**, *269*, 811-826.
48. Koo, H. S.; Crothers, D. M. *Proc Natl Acad Sci U. S. A.* **1988**, *85*, 1763-1767.

Chapter 4

Lattice- and Sequence-Dependent Binding of Mg^{2+} in the Crystal Structure of a B-DNA Dodecamer

Martin Egli[1] and Valentina Tereshko[2]

[1]Department of Biological Sciences, Vanderbilt University,
Nashville, TN 37235
[2]Department of Cellular Biochemistry and Biophysics, Memorial Sloan-
Kettering Cancer Center, New York, NY 10021

Crystal structures of nucleic acids often reveal multiple metal ions. Whether a certain ion is specific to the structural motif or the result of the crystal lattice and/or crystallization conditions is usually not obvious. Five Mg^{2+} ions per asymmetric unit were observed in the crystal structure of the B-form DNA duplex with sequence CGCGAATTCGCG. One of them binds at a GpC step, adjacent to a kink into the major groove near one end of the duplex. Two others link phosphate groups across the minor groove, resulting in a marked narrowing at one border of the AATT-tract. However, none of the ions binds to only a single duplex. Instead they mediate contacts between symmetry-related DNA dodecamers. Moreover, all Mg^{2+} ions exhibit either inner- or outer-sphere coordination to phosphate groups. Although some of the Mg^{2+} ions accentuate conformational features of the DNA duplex in the lattice, their particular locations and coordination modes indicate that ion binding to the dodecamer is primarily a consequence of the crystal packing and, to a lesser degree, of the sequence.

87

Metal ions play important roles in the stability of DNA (*1*) and RNA (*2, 3*), DNA conformational transitions (*4*) and phosphodiester cleavage and ligation reactions catalyzed by nucleic acids (*5*). The binding modes of ions can be grouped broadly into two categories: Diffuse and site-binding (*6*). Both can serve specific functions in the structural organization of nucleic acids and RNA catalysis. However, only site-bound or localized metal ions can be observed in X-ray crystal structures.

Many among the recently determined RNA crystal structures have revealed bound metal ions [(*7-11*), reviewed in (*12*)]. Similarly, improvements in the resolution of DNA oligonucleotide crystal structures have provided insight into the localizations of divalent (*13-17*) and monovalent metal cations (*18, 19*) [reviewed in (*20*)]. Which of these ions have to be considered an integral part of a certain RNA folding motif or a DNA duplex may often be difficult to determine. For example, analysis of the effects of monovalent cations on the stability of a 58-nucleotide fragment from *E. coli* 23S rRNA showed that K^+ was most effective in terms of stabilizing the RNA tertiary structure and that the increased stability was due to a single K^+ ion (*3*). A subsequent crystal structure of the fragment led to the identification of this ion and the precise mode of tertiary structure stabilization (*10*). The structure of the internal loop E fragment of 5S rRNA is known to be highly sensitive to the concentration of divalent metal ions and in the high-resolution crystal structure five Mg^{2+} ions were found to coordinate to the loop E major groove (*7*). However, it appears that only one of them may be specific to the loop E and hence of biological importance (*21*).

It has been correctly pointed out that the more complex tertiary and quaternary structure of RNA compared with (duplex) DNA leads to a more versatile role of metal ions, particularly monovalent ions, in the stabilization of RNA folding (*17*). Oligodeoxynucleotide duplexes constitute a relatively simple class of structural motifs despite the existence of right-handed and left-handed geometries. Because phosphate groups lie on the outside of the double helix a large portion of the helical surface displays strong electronegative potential. Most of the localized metal ions in crystal structures of nucleic acid duplexes simply relieve electrostatic repulsion between phosphates from adjacent strands and thus provide the mortar between the DNA bricks. When analyzing the effects of site-bound or localized metal ions on DNA duplex structure, it is obviously important to differentiate between sites that are occupied as a result of packing interactions and those that may constitute intrinsic binding pockets, potentially resulting in specific ion-mediated conformational properties of DNA. Another variable that has to be taken into account when judging the specificity of metal ion binding is the often high concentrations of cations used for crystallizing nucleic acid compared with their levels *in vivo*.

The B-form DNA oligonucleotide with sequence CGCGAATTCGCG, the so-called Dickerson-Drew dodecamer (*22*) (DDD), provides an excellent example for analyzing the potential effects of a particular class of divalent metal ions (Mg^{2+}) on the conformation of a double helical fragment. The structure of the Mg form has been determined at atomic resolution [max. 0.95 Å; ref (*23*)] and five Mg^{2+} ions could be located per crystallographic asymmetric unit (*13, 14*). The dodecamer duplex exhibits an asymmetric kink into the major groove that is associated with a bound Mg^{2+} hexahydrate and two additional Mg^{2+} ions were observed to coordinate to the phosphate backbone outside the groove adjacent to a further hallmark of the DDD, the narrow AATT portion of the minor groove. In the present chapter, we review the coordination modes of Mg^{2+} to the DDD and examine their possible influence on its structure in the orthorhombic crystal lattice. As the reader will see, a further analysis of the best studied DNA duplex structure from the point of view of DNA-cation interactions demonstrates that, although bound Mg^{2+} ions can modulate the conformation of the DDD, metal ion coordination is primarily determined by crystal packing and sequence.

The Ionic Environment of the DDD in the Mg Form

The crystal structure of the DDD at 1.1 Å resolution furnished the locations of five Mg^{2+} ions in the asymmetric unit (*13*). In the crystal lattice, each duplex is surrounded by 13 divalent ions as a result of the packing interactions in the orthorhombic space group (*14*). The Mg^{2+} environment for an individual dodecamer with the associated packing contacts is depicted in Plate 1. The five crystallographically independent ions are termed Mg1 to Mg5 and DNA residues in the first strand are numbered 1 to 12 and those in the second are numbered 13 to 24. Mg1, Mg3 and Mg5 are hexahydrates and Mg2 and Mg4 are pentahydrates. The detailed coordination modes and geometries were described in reference (*14*). Mg1 is bound inside the major groove adjacent to one end of the duplex (Plates 1A, 2). It uses three of its water ligands to contact O6 and N7 and O6 of residues G2 and G22, respectively. The remaining water ligands are involved in hydrogen bonds to phosphate oxygens of residues A6 (O1P and O2P) and T7 (O2P) from a symmetry related duplex. Mg2 is engaged in contacts to three different DDDs in the crystal lattice. The inner-sphere contact is made to an O1P oxygen from a first duplex (residue T19). Two additional water-mediated contacts are made to two symmetry related DDDs and involve O1P oxygens from residues G12 and G24, respectively. Mg3 forms outer-sphere contacts to two neighboring duplexes. The ion interacts with both the O1P and O2P oxygen of residue G10 from the first duplex and with O1P of residue A18 in the second. Mg4 links three neighboring duplexes and is engaged in an inner-sphere interaction to O1P of A17 from a first duplex and outer-sphere interactions to both phosphate oxygens of C9 from a second duplex as well as to O2P of G24 from a third. By comparison Mg5 is exclusively involved in outer-sphere coordinations. It bridges two duplexes by forming contacts to O2P of residue C21 from one duplex and to O1P and O2P of residues A5 and A6, respectively, from a second. An additional outer-sphere contact is established to the 5'-hydroxyl group of residue C13.

The majority of contacts by Mg^{2+} ions in the lattice of the DDD are formed to phosphate groups. In fact Mg1 and Mg5 are the only ions that exhibit interactions to base (O6 and N7 of G) and sugar atoms (5'-OH), respectively, besides interactions to phosphates. This clearly demonstrates the importance of Mg^{2+}-phosphate interactions for lattice formation and stabilization. Our own work has yielded evidence that the Mg^{2+} concentration in crystallizations of the DDD correlates with the packing density and the resolution of X-ray diffraction data (13). Increased Mg^{2+} concentrations improve the diffraction limit, in turn making possible observation of divalent ions previously not visible in electron density maps based on lower resolution data or observation of partially occupied ions. Because Mg1 was also present in structures of the DDD at medium resolution (24), independent of the concentration of spermine used for growing crystals, we may assume that this particular site is occupied first in the Mg-from lattice of the DDD. Mg2 and Mg3 coordinate in close vicinity from each other and together bridge phosphate groups across the minor groove. The two ions were not observed in earlier structures at lower resolution and their occupancies may depend on the relative concentrations of Mg^{2+} and spermine in crystallizations of the DDD. Mg4 and Mg5 exhibit only partial occupancies even in the structure of the DDD at atomic resolution, based on crystals grown in the presence of relatively high concentrations of Mg^{2+} (typically > 25 mM). Therefore, it is reasonable to conclude that these sites may only become occupied as the Mg1, Mg2 and Mg3 ions are fully bound.

The Major Groove Mg^{2+} Ion

Mg1 bridges G2 and G22 from opposite strands in the major groove, adjacent to a distinct kink into that groove (Plates 1 to 3). In the 1.1 Å crystal structure of the DDD the kink amounted to ca. 11° (13). This asymmetric compression of the major groove near one end of the duplex is a long noted feature of the DDD (22) and has been referred to as a 'facultative bend' (25) or 'annealed kinking' (26) [briefly reviewed in ref. (27)]. The latter term referred to the smooth and continuous nature of the bending in the DDD, whereby an increase in roll appeared to lead to a cascade of alterations in propeller twisting that was propagated halfway down the helix (26). The issue of bending was later revisited and it was concluded that GC/AT junctions are inherently bendable and can adopt straight or bent conformations under the influence of crystal packing forces (28). The bend itself is produced by rolling one base pair over the next along their long axes such that the major groove gets compressed.

Structures of DDD duplexes studied under different conditions or in crystals grown from a range of conditions display various degrees of kinking. For example, the absence of a kink in the crystal structure of d(CGCGAATT^{Br}CGCG) at high concentrations of 2-methyl-pentane-2,4-diol (MPD) was attributed to the specific steric limitations as result of the additional bromine located at the edge of the major groove (26) (Plate 2B). Although the

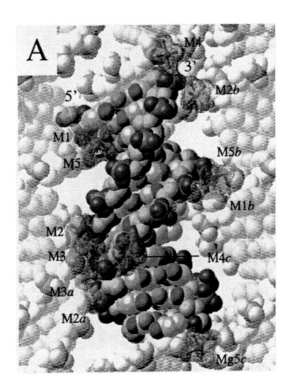

Plate 1. Mg^{2+} environment and packing contacts of the DDD duplex in the Mg form lattice. The dodecamer duplex viewed (**A**) into the minor groove, (**B**) into the major groove and (**C**) roughly along the z-direction of the orthorhombic unit cell. All molecules and metal cations are shown in van der Waals mode. Atoms of the DNA backbone are colored yellow, red and orange for carbon, oxygen and phosphorus, respectively. Nucleobase atoms are colored gray, pink and cyan for carbon, oxygen and nitrogen, respectively. Phosphate groups from neighboring duplexes are highlighted in magenta. Mg^{2+} ions are colored green and are labeled M1 through M5 with small letters in italic font designating individual symmetry mates. *(See page 7 of color insert.)*

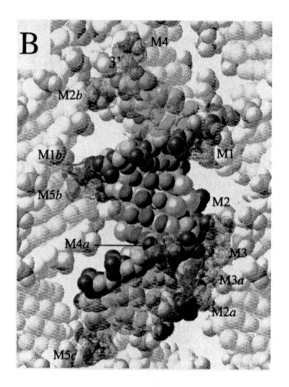

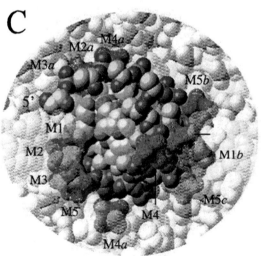

Plate 1. *Continued. (See page 7 of color insert.)*

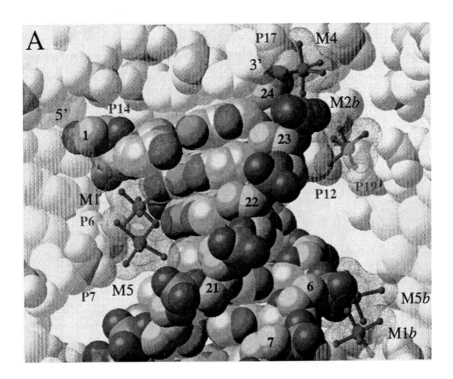

Plate 2. Close-up views of the top half of the DDD duplex (base pairs C1:G24 to A6:T19). **(A)** The duplex viewed across the major and minor grooves, illustrating the kink into the major groove and Mg1 bound at the G2pC3 (C22pG23) step. The color scheme for atoms is identical to that in plate 1 and Mg^{2+} ions and selected residues as well as phosphate groups from symmetry related duplexes are labeled. **(B)** Superposition of the top halves of DDD duplexes in selected structures: Nucleic acid database [NDB (*40*)] code BD0007 [(*13*), green], BD0005 [(*32*), cyan], BDL001 [(*22*), gray] and BDLB04 [(*26*), pink], demonstrating the similar degrees of kinking into the major groove in the first three structures and the absence of a kink in the structure of the brominated DDD [d(CGCGAATT^{Br}CGCG)]₂. *(See page 8 of color insert.)*

94

B

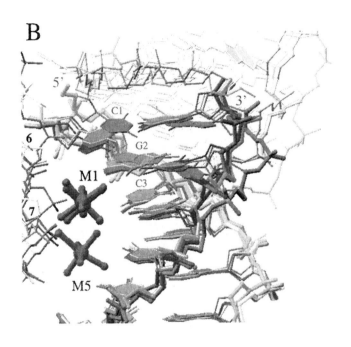

Plate 2. *Continued. (See page 8 of color insert.)*

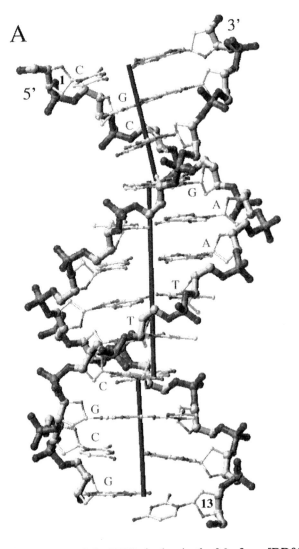

Plate 3. Helical geometry of the DDD duplex in the Mg form [BD0007 (*13*)].
(**A**) The duplex viewed into the minor groove, roughly along the molecular
dyad. Helical axes for the CGCG tetramers at both ends and the central hexamer
GAATTC were calculated with the program CURVES (*41*) and are depicted in
blue. The color scheme for atoms is identical to that in plate 1 and residues of
strand 1 are labeled. The drawing illustrates the asymmetric kink of the DDD
duplex into the major groove, based in part on a positive roll at the C3pG4
(C21pG22) step. (**B**) Superposition of the top (pink) and bottom (cyan) halves
of the DDD duplex. Accordingly, the central hexamer displays almost perfect
twofold rotational symmetry, while the kink near one end compresses the major
groove and results in deviating geometries of the two 'G-tracts'. The DDD
duplex in the orthorhombic lattice can be thought of as composed of a trimer and
a nonamer duplex that are themselves straight but exhibit a ca. 11° kink at their
interface. *(See page 9 of color insert.)*

B

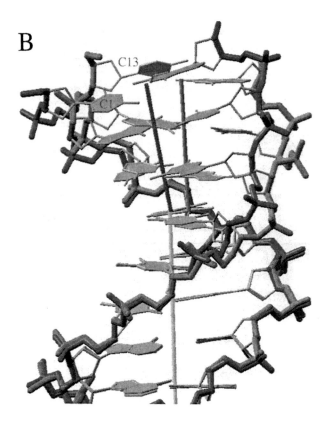

Plate 3. *Continued. (See page 9 of color insert.)*

exact origins of the asymmetric kink remained unclear, the tendency to roll and bend must be consequence of the particular sequence context (28). Both packing forces (22) and kinetics of crystallization (26) were invoked as explanations for the asymmetric kink. The observation that the [d(CGCGAATT^{Br}CGCG)]₂ duplex was straight suggests that a single bromine can overcome the local packing forces (intermolecular hydrogen bonds) in DDD crystals. It is noteworthy that both straight and bent DDDs can apparently be accommodated in the orthorhombic crystal lattice (28).

In the original structures of the DDD with resolutions between 2.3 and 3.0 Å Mg1 had not been observed (22, 25, 26). In some reports, a spermine molecule was described to cross the major groove at the site of the greatest bending (26, 29), but such a location of the polyamine was never confirmed and in the structures at high resolutions, spermine was not found in that region either. However, the presence of Mg1 near the site of the kink in recent crystal structures of the DDD provided new support to the idea that a cation could somehow promote the kink (20). Mg1 was initially observed in crystal structures of the DDD with base pair mismatches or drugs bound [i.e. (30)], perhaps as a result of the improved quality of electron density maps due to the advent of low-temperature data collection. This Mg^{2+} ion was also present in DDD structures at medium resolution (24, 31) and high occupancy of the site is apparently independent of the specific ratios between Mg^{2+} and spermine used for growing crystals. However, other divalent metal cations such as Ca^{2+} do not enter the site – Ca^{2+} actually coordinates in the minor groove of the DDD (14) – and alkali metal cations may not be able to replace Mg^{2+} unlike at a site inside the major groove of an A-form decamer duplex (19). In terms of the role of Mg^{2+} in inducing or stabilizing bending, a helix with an asymmetric kink and Mg^{2+} bound nearby may crystallize easily, in any case more easily than a symmetrically kinked duplex with ions bound near both ends or a straight duplex (26). However, one should not discount a potentially crucial role of packing forces in either inducing or enhancing the kink. As described above, Mg1 does not just engage in contacts to a singe duplex, but is inserted between the major groove and phosphates of the C1-G12 strand from a second duplex (Plates 1, 2). Mg5 is bound in close vicinity but only forms a hydrogen bond to a phosphate oxygen of C21 in an outer-sphere fashion and none of the ligand waters are shared between the two ions. This close approach between phosphates and a major groove involving two duplexes is not observed at the other end of the DDD (Plate 4). Because the environments of the duplex ends are different in the orthorhombic crystal and the asymmetric kink occurs near the end that is involved in tight Mg^{2+}-stabilized inter-duplex contacts, packing forces could be enhancing and stabilizing the particular conformation of the DDD in the Mg form.

The DDD duplex does not exhibit a kink at the opposite end (C12:G13; Plate 4). The helical axes of the [d(CGCG)]₂ and the central [d(GAATTC)]₂ duplex portions are virtually parallel (Plate 3). The drawings of the two halves of the DDD duplex and the surrounding duplexes depicted in Plates 2 and 4 illustrate that the packing around the C1:G24 end of the duplex is tighter than that around the G12:C13 end. The superposition of DDD duplexes from different structures reveals that C13 enjoys a certain degree of freedom (Plate 4B), clearly supporting the notion that packing there is somewhat less tight. Although certain features of the environment of the two DDD duplex ends are quite similar, such as for example the way they overlap with the terminal base pairs from duplexes above and below in the crystallographic z-direction of the lattice (25), the packing deviate to a certain extent. In some structures of the DDD, fragments of spermine resided near the major groove at a site that is equivalent to the location of Mg1 at the opposite end of the duplex (32, 33). However, the presence of the cation there appears to have no conformational consequences (Plate 4B). In all DDD structures reported to date this end of the molecule is straight. Therefore, it is reasonable to suspect that alternative influences by the packing might play a role in bringing about the particular geometry of the DDD duplex observed in virtually all of these crystals.

The Minor Groove Mg^{2+} Ions

Mg2 and Mg3 are located at the periphery of the narrow central minor groove of the DDD duplex (13). Thus, they link phosphates from residues T19 and G10 across the groove (Plate 5). A symmetry-related Mg^{2+} pair, M2a and M3a (Plate 5), also gets to lie near the minor groove, interacting with phosphate oxygens of residues G12 and A18, respectively. None of these ions establishes any interaction to a base atom at the floor of the groove and the coordination spheres are only involving phosphate groups. Interestingly, the groove at the site of the Mg^{2+} bridge between P10 and P19 is markedly narrower relative to DDD duplexes in structures that did not contain these ions (14) (Plate 6). In fact, the groove at that site is more compressed than in any other structure of the DDD, the actual reduction in width amounting to around 1 Å relative to the duplexes included in the comparison shown in Plate 6.

The crystallographic data clearly demonstrated that divalent metal cations, even in cases where they are located at the periphery of a groove, can modulate the groove width by relieving electrostatic repulsion between closely spaced phosphate groups from opposite strands. Apparently, this can occur in sections of the minor groove that already exhibit sequence-dependent narrowing, as is the case for the AATT stretch in the DDD (Plate 6). By contrast, coordination of alkali metal cations inside the central portion of the DDD minor groove has not resulted in any observable changes of groove width according to all available crystallographic data (18, 20, 31, 32). However, in a recent crystal structure of a stilbenediether-capped DNA hexamer duplex with an A$_4$:T$_4$ tract, a Mg^{2+} hexahydrate complex was located inside the A-tract minor groove (34). The Mg^{2+} ion interacts with O2 and N3 atoms of a thymine and two adenines, respectively, and forms further contacts to 4'-oxygen atoms of two residues. It is noteworthy that in this structure, unlike in the case of Mg1 in the DDD major

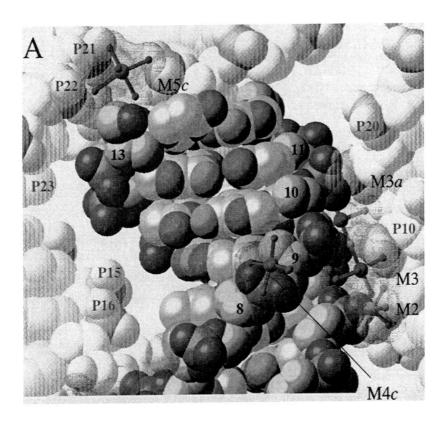

Plate 4. Close-up views of the bottom half of the DDD duplex (base pairs C13:G12 to A18:T7). **(A)** The duplex viewed across the major and minor grooves, illustrating the absence of a kink at the site equivalent to that in the top half of the duplex (plate 2) and the absence of Mg^{2+} in that portion of the major groove. The color scheme for atoms is identical to that in plate 1 and Mg^{2+} ions and selected residues as well as phosphate groups from symmetry related duplexes are labeled. **(B)** Superposition of the bottom halves of DDD duplexes in selected structures: Nucleic acid database [NDB (40)] code BD0007 [(13), green], BD0005 [(32), cyan], BDL001 [(22), gray] and BDLB04 [(26), pink], demonstrating that all four duplexes are straight and that binding of a spermine molecule in one case [only 6 atoms were observed (32)] does not affect the helical geometry. The drawing also illustrates the conformational flexibility of the terminal base pair C13:G12, resulting in effective unstacking of C13 in some duplexes. This provides an indication that the constraints due to packing are considerably different for the two DDD duplex ends, the packing around the C13:G12 base pair apparently being less tight. *(See page 10 of color insert.)*

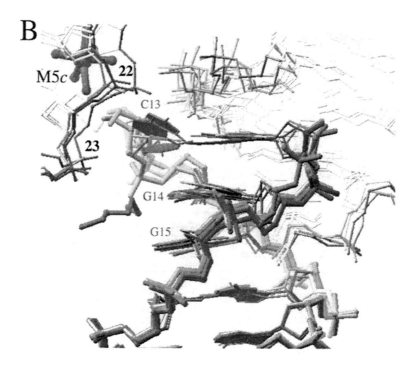

Plate 4. *Continued. (See page 10 of color insert.)*

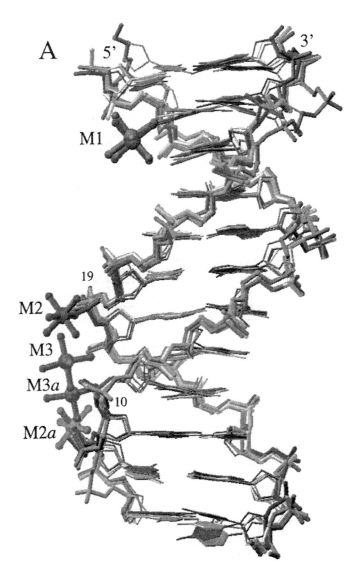

A 5' 3'

M1

19

M2

M3

M3a

10

M2a

Plate 5. Mg^{2+} coordination in the minor groove of the DDD. (**A**) Superposition of DDD duplexes based on structures BD0007 [(*13*), green], BD0005 [(*32*), cyan], BDL001 [(*22*), gray] and BDLB04 [(*26*), pink], indicating the slight local narrowing of the minor groove at the site of Mg^{2+} coordination in the BD0007 duplex. (**B**) In the structure of the DDD duplex at 1.1 Å resolution, a tandem of Mg^{2+} ions (M2 and M3) crosses the minor groove at the periphery, linking phosphate groups from opposite strands (*13*). Two symmetry-related Mg^{2+} ions (M2a, M3a) get to lie in the vicinity of the minor groove near the G12:C13 end of the same duplex. The color scheme for atoms is identical to that in plate 1 and selected DNA residues are labeled. *(See page 11 of color insert.)*

B

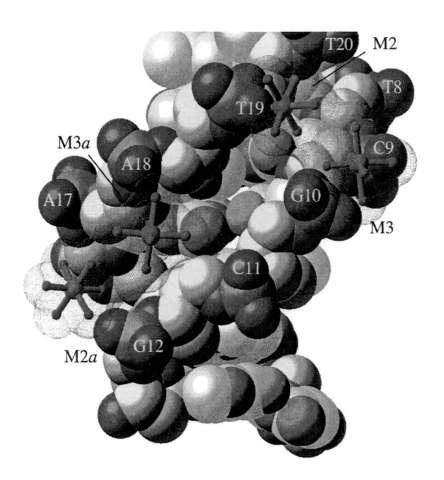

Plate 5. *Continued. (See page 11 of color insert.)*

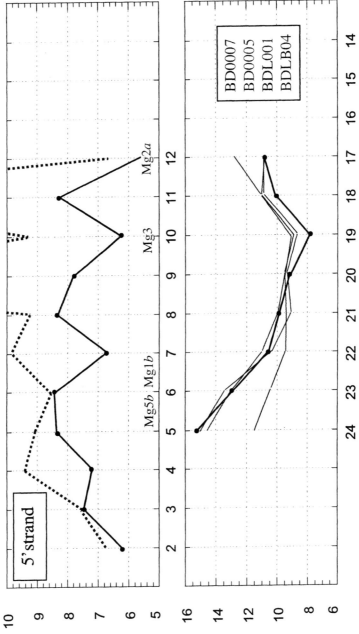

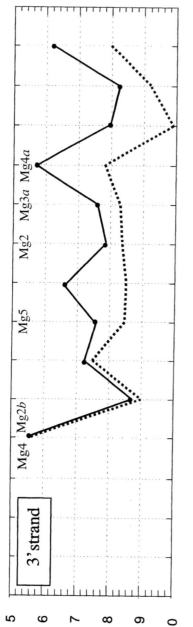

Plate 6. Intra- (defining minor groove width) and inter-duplex P⋯P distances in Å in the high-resolution crystal structure of the DDD (*13*) and Mg²⁺ ions stabilizing closely spaced phosphate pairs in the orthorhombic lattice. The graphs at the top and at the bottom depict inter-duplex distances between phosphorus atoms in strands C1-G12 and C13-G24, respectively, and phosphorus atoms from neighboring duplexes. The closest contacts for Ps in each strand are connected by a solid line and the second closest contacts are connected with a dashed line. The graph in the middle depicts minor groove widths in four DDDs: BD0007 [(*13*), green], BD0005 [(*32*), cyan], BDL001 [(*22*), gray] and BDLB04 [(*26*), pink] and illustrates the narrower minor groove of the DDD as a result of the Mg2-Mg3 bridge across the groove in the BD0007 structure. The coordination sites of Mg²⁺ ions are indicated.

(See page 12 of color insert.)

groove or Mg2 and Mg3 in the DDD minor groove, the Mg^{2+} ion binds to only one duplex. As a result, the A-tract minor groove is narrower by almost 1 Å compared with a second molecule in the asymmetric unit which does not display Mg^{2+} coordination at that site.

Stabilization of Short Intermolecular P···P Contacts by Mg^{2+}

What determines the locations of Mg^{2+} ions in the orthorhombic lattice of the DDD? As pointed out in the introduction, X-ray crystallography even at high resolution can only shed light on the whereabouts of site-bound ions whereas delocalized ones remain unaccounted for. So an answer to the question regarding factors that affect the preference for particular sites is necessarily incomplete as only a subset of cations is commonly included in such an analysis. With the exception of a handful of cases (35-38), the cations retrieved in crystal structures of oligonucleotides normally do not account for a complete neutralization of the negatively charged phosphate groups.

To answer the above question in the case of the five Mg^{2+} ions per asymmetric unit in DDD crystals, let us first look at the coordination spheres of these ions. All of them feature at least one interaction to a phosphate group. For two among the five, this interaction is of the inner sphere mode. Moreover, none of the Mg^{2+} ions binds only to a single duplex; they all establish contacts to at least two adjacent DDD molecules. Next, let us analyze the inter- and intra-duplex (inter-strand) distances between phosphorus atoms. These are graphically summarized in Plate 6. The graphs at the top and at the bottom show distances between phosphorus atoms in strands 1 and 2, respectively, and phosphorus atoms from symmetry-related duplexes. In each case, the two closest P···P contacts in Å are depicted. The graph in the middle shows distances between phosphorus atoms from opposite strands across the minor groove. Basically, these distances provide a measure for the minor groove width.

There are eleven inter-duplex P···P distances below 7 Å (Plate 6, top and bottom). If we account for symmetry only seven unique interactions remain. The shortest intra-duplex P···P distance is 7.8 Å (between P10 and P19; Plate 6, middle) and corresponds to the narrowest site of the minor groove. Except for the close contact between P2 and P14# from two different DDD molecules (# indicates a symmetry-related duplex), each of these eight close P···P contacts in the lattice is stabilized by a Mg^{2+} ion (P10 and P19 are bridged by two ions across the minor groove as described in the previous section). The P2···P14# contact is visible in Plate 2A (upper left), although P14# is partly obscured in the drawing. Inspection of the surroundings of the P2···P14# pair reveals that they face a solvent channel and are bridged by a single water molecule. Interestingly, Mg1 and Mg5c bracket the two phosphates, although neither ion establishes a direct (inner- or outer-sphere) coordination to either of the two phosphates. The distance between Mg1 and P2 is 9.4 Å (Plate 2A) and Mg5c is equidistant from P2 and P14# (8.9 Å; Plate 4A).

From this analysis it appears that the basic role of ordered Mg^{2+} ions in DDD crystals is the alleviation of potentially destabilizing electrostatic repulsions between closely spaced phosphate groups, residing either in two different duplexes or within a duplex. Therefore, the particular relative orientation of DDD duplexes in the three-dimensional lattice emerges as a major determinant of the arrangement of Mg^{2+} ions around an individual oligonucleotide duplex. In turn, ions trapped by the lattice can affect the local geometry of the DDD duplex, as illustrated by the asymmetric kink into the major groove (Plates 2, 3) and the local contraction of the minor groove (Plate 5).

Conclusions

The organization of divalent metal ions around the DDD duplex in the crystal structure of the Mg form is chiefly determined by packing interactions. There are two main reasons for the importance of the crystal lattice, or, in other words, the relative orientations of DNA duplexes, in controlling Mg^{2+} ion localization. One is the absence of complex tertiary and quaternary structural elements in double helical DNA (17). Mg^{2+} ions can interact with the backbones or the grooves or both. However, because the negatively charged phosphate groups are exposed on the surface of the DNA duplex, they dominate the coordination spheres of Mg^{2+} ions. Thus, none of the five ions in the DDD structure exhibits binding to base atoms only; each Mg^{2+} ion forms at least one contact to a phosphate group, either via the inner- or outer sphere mode. The second reason is related to the relatively high enthalpy of hydration of the magnesium hexahydrate ion. Two of the Mg^{2+} ions in the crystal structure of the Mg form of the DDD show a single inner-sphere interaction to a DNA phosphate group (Mg2 and Mg4). Both are engaged in at least one additional water-mediated interaction to a phosphate group from a symmetry-related DDD. They are representative of a more general observation with regards to the coordination preferences of Mg^{2+}: There is not a single case of a DNA crystal structure where Mg^{2+} was observed to form a direct (inner-sphere) contact to just one DNA molecule. The water ligands of such ions are always hydrogen bonded to one or more phosphate groups from adjacent DNAs. Apparently, a single phosphate group cannot compensate for the loss of a water or a binding site located within a single DNA molecule is no match for those that involve two or three neighboring molecules offering geometrically more optimally oriented phosphate groups that fit the rigid octahedral coordination geometry of Mg^{2+}. In principle, one could imagine a Mg^{2+} ion being bound between two adjacent phosphate groups from the same backbone. However, in the structure of the DDD such a coordination mode is not present. It is likely that the spacing of duplexes in DNA crystal structures is such that an ion bound to phosphates from one backbone would get to lie in close vicinity of a backbone from a symmetry-related duplex. An alternative explanation for the absence of the above coordination mode is that, perhaps, an ion bound only to adjacent phosphate groups of one duplex and facing an open solvent-filled region in the crystal exhibits high mobility and may then be 'invisible' in electron density maps even at high resolution.

There are two instances in the crystal structure of the DDD where Mg^{2+} coordination is associated with a change in DNA conformation. The first is the kink into the major groove adjacent to Mg1 (Plates 2 and 3) and the second is the further narrowing of the minor groove as a result of the Mg2-Mg3 bridge in the AATT portion of the DDD (Plate 5). These cases demonstrate the influence of sequence in directing Mg^{2+} binding in conjunction with that of the packing described above. Thus, sequence-dependent narrowing of the minor groove in the A-tract of the DDD allows bridging by the Mg^{2+} tandem in the first place; ion binding then brings about a further contraction. Regarding the coordination of Mg1 at the G2pC3 step, it is certainly possible that the conservation of the kink in most crystal structures of the DDD is a result of the kinetics of crystallization. However, this still leaves the DNA sequence as the basic impetus for the preferred binding of Mg^{2+} at the GpC step. Again, packing plays an important role as well because Mg1 links the DDD duplex at the site of the kink to the backbone of an adjacent duplex (Plate 2). At the chemically equivalent G14pC15 step near the other end of the duplex no ion is found (Plate 4). The absence of Mg^{2+} at that site is likely related to the looser packing in the DDD crystal there, with the major groove being farther removed from the backbones of neighboring DDDs.

Regarding the order of events that result in sequence-dependent conformational changes associated with binding of divalent metal ions, the recent analysis of Mn^{2+} coordination in the crystal structure of the nucleosome core particle (NCP) has been very instructive (39). The overall bend and topology of the 147 base pair DNA duplex in the NCP are a consequence of the histone proteins in its core. Packing forces are necessarily less important in directing ion binding in the case of the NCP. Moreover, the softer Mn^{2+} ion compared with Mg^{2+} exhibits more inner-sphere coordination to N7 of guanine and is therefore more commonly associated with a single DNA molecule rather than being trapped between neighboring molecules. In the structure of the NCP, GpG and GpC steps harboring Mn^{2+} ions display characteristic roll, slide and shift parameters (39), demonstrating the importance of the duplex DNA conformation imposed by the nucleosome core on divalent metal ion binding. Accordingly, DNA conformation dictates metal ion binding and not *vice versa*.

In summary, DNA sequence (determines DNA conformation and hence) directs metal ion binding and, in turn, bound cations may modulate DNA conformation. In crystal structures of oligonucleotide DNA duplexes, such as the DDD discussed here, the lattice assumes a role in governing Mg^{2+} coordination that is more important than sequence.

108

Acknowledgment

This work is supported by the National Institutes of Health (grant GM55237 to M.E.).

Literature Cited

1. Felsenfeld, G.; Rich, A. *Biochim. Biophys. Acta* **1957**, *26*, 457-468.
2. Misra, V.K.; Draper, D.E. *Biopolymers, Nucl. Acid Sci.* **1998**, *48*, 113-135.
3. Shiman, R.; Draper, D.E. *J. Mol. Biol.* **2000**, *302*, 79-91.
4. Behe, M.J., Felsenfeld, G. *Proc. Natl. Acad. Sci. U.S.A.* **1981**, *78*, 1619-1623.
5. Fedor, M.J. *Curr. Opin. Struct. Biol.* **2002**, *12*, 289-295.
6. Misra, V.K.; Draper, D.E. *Proc. Natl. Acad. Sci. U.S.A.* **2001**, *98*, 12455-12461.
7. Correll, C.C.; Freeborn, B.; Moore, P.B.; Steitz, T.A. *Cell* **1997**, *91*, 705-712.
8. Cate, J.H.; Hanna, R.L.; Doudna, J.A. *Nature Struct. Biol.* **1997**, *4*, 553-558.
9. Basu, S.; Rambo, R.P.; Strauss-Soukup, J.; Cate, J.H.; Ferré-D'Amaré, A. R.; Strobel, S.A. Doudna, J.A. *Nature Struct. Biol.* **1998**, *5*, 986-992.
10. Conn, G.L.; Gittis, A.G.; Lattman, E.E.; Misra, V.K.; Draper, D. E. *J. Mol. Biol.* **2001**, *318*, 963-973.
11. Egli, M.; Minasov, G.; Su, L.; Rich, A. *Proc. Natl. Acad. Sci. U.S.A.* **2002**, *99*, 4302-4307.
12. Egli, M.; Minasov, G. In *Ribozymes, Biochemistry and Biotechnology*; Krupp, G.; Gaur, R. Eds.; Eaton Publishing: Natick, MA, 2000; pp 315-349.
13. Tereshko, V.; Minasov, G.; Egli, M. *J. Am. Chem. Soc.* **1999**, *121*, 470-471.
14. Minasov, G.; Tereshko, V.; Egli, M. *J. Mol. Biol.* **1999**, *291*, 83-99.
15. Soler-Lopez, M.; Malinina, L.; Liu, J.; Huynh-Dinh, T.; Subirana, J.A *J. Biol. Chem.* **1999**, *274* 23683-23686.
16. Kielkopf, C.L.; Ding, S.; Kuhn, P.; Rees, DC. *J. Mol. Biol.* **2000**, *296*, 787-801.
17. Chiu, T.K.; Dickerson, R.E. *J. Mol. Biol.* **2000**, *301* 915-945.
18. Tereshko, V.; Minasov, G.; Egli, M. *J. Am. Chem. Soc.* **1999**, *121*, 3590-3595.
19. Tereshko, V.; Wilds, C.J.; Minasov, G.; Prakash, T.P.; Maier, M.A.; Howard, A.; Wawrzak, Z.; Manoharan, M.; Egli, M. *Nucleic Acids Res.* **2001**, *29*, 1208-1215.
20. Egli, M. *Chem. Biol.* **2002**, *9*, 277-286.
21. Serra, M.J.; Baird, J.D.; Dale, T.; Fey, B.L.; Retatagos, K.; Westhof, E. *RNA* **2002**, *8*, 307-323.
22. Wing, R. Drew, H.; Takano, T.; Broka, C.; Takano, S.; Itakura, K.; Dickerson, R.E. *Nature* **1980**, *287*, 755-758.

23. Egli, M.; Tereshko, V.; Teplova, M.; Minasov, G.; Joachimiak, A.; Sanishvili, R.; Weeks, C.M.; Miller, R.; Maier, M.A.; An, H.; Cook P.D.; Manoharan, M. *Biopolymers, Nucl. Acid Sci.* **2000**, *48*, 234-252.

24. Berger, I.; Tereshko, V.; Ikeda, H.; Marquez V.E.; Egli, M. *Nucleic Acids Res.* **1998**, *26*, 2473-2480.

25. Drew, H.R.; Wing, R.M.; Takano, T.; Broka, C.; Tanaka, S.; Itakura, K.; Dickerson, R.E. *Proc. Natl. Acad. Sci. U.S.A.* **1981**, *78*, 2179-2183.

26. Fratini, A.V.; Kopka, M.L.; Drew, H.R.; Dickerson, R.E. *J. Biol. Chem.* **1982**, *257*, 14686-14705.

27. Allemann, R.K.; Egli, M. *Chem. Biol.* **1997**, *4*, 643-650.

28. Dickerson, R.E.; Goodsell, D.S.; Neidle, S. *Proc. Natl. Acad. Sci. U.S.A.* **1994**, *91*, 3579-3583.

29. Drew, H.R.; Dickerson, R.E. *J. Mol. Biol.* **1981**, *151*, 535-556.

30. Quintana, J.R.; Lipanov, A.A.; Dickerson, R.E.; *Biochemistry* **1991**, *30*, 10294-10306.

31. Shui, X.; McFail-Isom, L.; Hu, G.G.; Williams, L.D. *Biochemistry* **1998**, *37*, 8341-8355.

32. Shui, X.; Sines, C.C.; McFail-Isom, L.; VanDerveer, D.; Williams, L.D. *Biochemistry* **1998**, *37*, 16877-16887.

33. Minasov, G.; Teplova, M.; Nielsen, P.; Wengel J.; Egli, M. *Biochemistry* **2000**, *39*, 3525-3532.

34. Egli, M.; Tereshko, V.; Mushudov, G.N. Sanishvili, R.; Liu, X.; Lewis, F.D. *J. Am. Chem. Soc.* **2003**, *125*, in press.

35. Wang, A.H.-J. Quigley, G.J.; Kolpak, F.J.; Crawford, J.L.; van Boom, J.H. ; van der Marel, G.; Rich, A. *Nature* **1979**, *282*, 680-686.

36. Gessner, R.V.; Frederick, C.A.; Quigley, G.J.; Rich, A.; Wang, A.H.-J. *J. Biol. Chem.* **1989**, *264*, 7921-7935.

37. Bancroft, D. Williams, L.D. Rich, A. Egli, M. *Biochemistry* **1994**, *33*, 1073-1086.

38. Kielkopf, C.L.; Ding, S.; Kuhn, P.; Rees, D.C. *J. Mol. Biol.* **2000**, *296*, 787-801.

39. Davey, C.A.; Richmond, T.J. *Proc. Natl. Acad. Sci. U.S.A.* **2002**, *99*, 11169-11174.

40. Web address: http://ndbserver.rutgers.edu.

41. a) Lavery, R.; Sklenar, J. *J. Biomol. Struct. Dyn.* **1988**, *6*, 63-91; b) Lavery, R.; Sklenar, H. *J. Biomol. Struct. Dyn.* **1989**, *7*, 655-667.

Chapter 5

Phosphate Crowding and DNA Bending

P. R. Hardwidge[1,4], Y.-P. Pang[2], J. M. Zimmerman[1], M. Vaghefi[3],
R. Hogrefe[3], and L. J. Maher, III[1,5,*]

[1]Department of Biochemistry and Molecular Biology, Mayo Clinic,
Rochester, MN 55905
[2]Department of Molecular Pharmacology and Experimental Therapeutics,
Mayo Clinic, Rochester, MN 55905
[3]TriLink BioTechnologies, 6310 Nancy Ridge Drive, Suite 101, San
Diego, CA 92121
[4]Current address: Biotechnology Laboratory, #237–6174 University
Boulevard, University of British Columbia, Vancouver, British Columbia
V6T 1Z3, Canada
[5]Current address: 200 First Street SW, Rochester, MN 55905
(maher@mayo.edu, telephone: 507–284–9041, fax: 507–284–2053)

Although the structure of DNA has been known for fifty
years, it remains unclear how electrostatic effects influence
DNA conformation. Here we ask whether there should be *any*
electrostatic penalty for DNA bending. Could it be that the
electrostatic cost of phosphate compression on the inner face
of bent DNA is compensated by favorable phosphate dilution
on the outer face? We address this issue by posing two simple
questions. First, does DNA bending result in a net crowding
of charged phosphate diester groups? Second, does artificial
crowding of charged phosphate groups result in DNA
bending? We begin by applying electrostatic energy
calculations to molecular models of DNA in a linear or bent
conformation. We show that i) there is indeed a penalty in
electrostatic interaction energy for bending the DNA double
helix, and ii) this penalty for bending is due to crowding of
phosphates separated over the range of 4-17 Å, and is
incompletely compensated by phosphate dilution on the outer
(convex) DNA surface. We then describe a DNA bending
experiment in which the addition of two negative charges by
site-specific phosphonylation of the DNA major groove is
shown to induce a detectable bend away from these appended
charges.

111

Introduction

Because of its densely-charged polyanionic backbones, electrostatic effects are expected to govern the physical properties of double-stranded DNA. Examples of these effects include counterion condensation (*1-3*) and the ion-exchange character of equilibria involving DNA binding proteins (*4-7*).

We have been interested in testing the hypothesis that local DNA bending can be induced by disturbing local charge distributions in the DNA grooves (*8, 9*). According to a "charge collapse" model, asymmetrically unbalancing local interphosphate repulsions on one face of the DNA double helix can cause the collapse of one of the DNA grooves, resulting in axial bending (*10, 11*). To test this idea, our previous experiments have focused on charge neutralization across the DNA minor groove, either by direct phosphate neutralization [methylphosphonate substitution; (*12-17*)] or by tethering individual ammonium ions to multiple nucleosides across a minor groove (*18-20*). We have also performed experiments with DNA-binding peptides that hold charged amino acids near DNA (*21-27*). Based on our results, we have proposed that the extent of DNA bending can be controlled by the valance and density of charges positioned in the DNA major groove. This is important because proteins often hold charged amino acids near DNA. Metal ions (monovalent and divalent) as well as polyamines (multivalent) are also natural DNA ligands. The potential distortion of DNA by such ions is of great current interest (*28-34*).

In a particularly intriguing report, molecular modeling experiments were used to predict the manner in which unbalanced interphosphate repulsions would alter DNA structure (*35*). This work suggested that local reduction of the dielectric constant in a DNA groove upon displacement of water by a protein domain would enhance interphosphate repulsions across this groove, giving rise to dramatic alterations in DNA structure. We have reasoned that similar effects should result from increasing the total negative charge in a DNA groove without altering the solvent dielectric constant.

Our experimental strategy tests the hypothesis that DNA charge-collapse will cause bending toward cations and away from anions. By covalently appending the charged group at a specific location within a synthetic DNA duplex, DNA bending can then be monitored by a sensitive electrophoretic assay. We have now conceived and implemented studies where DNA shape is monitored upon site-specific tethering of an anionic function in the major groove.

In the present work, we begin by theoretically exploring the simple but profound question of whether the double-helical array of charged phosphate groups in DNA should make any unfavorable electrostatic contribution to DNA bending. In essence, is there net phosphate crowding upon DNA bending? This theoretical study considers the degree to which the unfavorable electrostatic consequences of phosphate crowding on the concave inner surface of bent DNA are compensated by the favorable electrostatics of phosphate dilution on its convex outer surface. We find that this electrostatic compensation is incomplete, with a net unfavorable contribution of phosphate electrostatics to DNA bending observed independent of the solvent dielectric constant in the model. We then experimentally study the DNA charge-collapse hypothesis by asymmetrically adding negative charges to a single position in the DNA major groove. Such an arrangement should be electrostatically unfavorable. Will axial bending result from enhanced phosphate repulsions on one DNA face? We develop an experimental system in which tethering of a phosphonate dianion in the DNA major groove is observed to induce axial bending away from the negative charges. Analysis of these experiments also suggests a possible role for tether character and location.

Materials and Methods

DNA electrostatic calculations

The three-dimensional structure of 71-bp segment of severely-bent DNA was extracted from the crystal structure of the nucleosome [pdb code 1aoi, (36)]. A corresponding model of the same 71-bp sequence was generated in a linear conformation with a helical repeat of 10.5 bp/turn using the biopolymer module of Insight II 97.0 software (Molecular Simulations, Inc.). Hydrogen atoms were added to the linear and bent DNA constructs using the EDIT module of the AMBER5.0 program. The energies of the DNA structures were then minimized using a positional constraint applied to all heavy atoms using the SANDER module of AMBER5.0 with the AMBER94 force field (37, 38). Thirty-five inner-phosphates were selected from the full bent DNA structure by visual inspection using the QUANTA 98 program (Molecular Simulations, Inc.). Five consecutive inner phosphates were selected and the following five consecutive outer phosphates were excluded for each helical turn of the bent DNA. The 35-outer-phosphate structure was similarly obtained. Phosphate crowding in various DNA geometries was assessed by calculations of the total electrostatic interaction energies of the linear/bent DNA structures and the inner/outer phosphate structures with a zero charge set on all atoms except for P, O1P, O2P, and O5', which have the AMBER94 RESP charges of 1.1659, -0.7761, -0.7761, and -0.4954, respectively.

Synthesis of DNA duplexes

Unmodified DNA oligonucleotides were synthesized by standard methods, and were purified by denaturing polyacrylamide gel electrophoresis and quantitated based on predicted molar extinction coefficients (*39*).

Modified oligonucleotides (e.g. the 36-mer, 5'-GAGC$_2$GCGT<u>X</u>TCGC-GCGGT$_5$GC$_2$GCT$_5$CG$_2$-3') were synthesized using a control pore glass solid support and standard phosphoramidite methods on an Expedite automatic DNA-synthesizer. "X" indicates a 4-(1,2,4-triazol-1-yl)-1-(β-D-2'-deoxyribofuranosyl)pyrimidin-2-one) substitution ("4-triazolyl-dU") obtained using commercial reagents from Glen Research (Sterling, VA). The support-bound oligonucleotide was treated with 2-aminoethylphosphonate (0.4 M) or 3-amino-1-propanol (0.4 M) and triethylamine (0.8 M) in methanol/water (3:2, 2.5 mL) at room temperature overnight. The mixture was then treated with concentrated ammonium hydroxide for 12 h at 55 °C, removed from the support and dried. A small amount of the original solid-support bound oligonucleotide was treated with hot ammonium hydroxide only and used as a reference. Both sample and control were analyzed by HPLC. Products were purified by denaturing gel electrophoresis and analyzed by mass spectrometry.

Oligonucleotides were radiolabeled by incubation with polynucleotide kinase (New England Biolabs, Beverly, MA) and [γ-^{32}P]-ATP prior to a chase with unlabeled ATP. Equal amounts of complementary oligonucleotides were then mixed in the presence of 200 mM NaCl and annealed by heating and slow cooling to produce duplexes for ligation between phasing arms.

Semi-synthetic phasing analysis of DNA bending

DNA bending analysis was performed as previously described (*40*), using a semi-synthetic approach to ligate 36-bp DNA duplexes with appropriate terminal overhangs between longer restriction fragments to generate a family of five ~250-bp trimolecular ligation products for each synthetic duplex to be tested. Each of the five products positions the synthetic sequence (whose shape is to be analyzed) at a different distance from a phased array of three A$_5$-tracts in the right arm of the construct. The 100-bp left arm is not intrinsically curved. Radiolabeled ligation products were resolved on 8% native polyacrylamide (1:29 bis-acrylamide:acrylamide) gels in 0.5x TBE buffer, at 10 V cm^{-1} at 22°C for 5 h. For each group of five phasing products, the gel mobility, μ, was measured in mm. For plotting purposes, this value was normalized by dividing by the average gel mobility of all of the five probes in the group, μ_{avg}. Curve-fitting to generate least-squares estimates of the amplitude and horizontal displacement of a cosine phasing function through the data was executed as described (*40*). Fitting of phasing data for a series of curved standards containing 0-3 phased A_5-tracts provided a linear relationship between amplitude of the phasing function and DNA curvature over the range 0° - 54° in the test duplex. This relationship was then used to deduce the apparent curvature of the synthetic DNA duplex where one A_5-tract was replaced by the sequence whose shape was to be tested.

Results and Discussion

Theory: electrostatics of phosphate distributions in bent DNA

In their theoretical treatment of DNA charge-collapse by asymmetric phosphate neutralization, Manning and co-workers used a two-dimensional ladder model of DNA to conclude that collapse and consequent axial bending should accompany unbalancing phosphate repulsions along one DNA strand (*11*). This treatment suggested that upon asymmetric charge neutralization, DNA bending toward the neutralized surface becomes spontaneous because the relaxation of interphosphate repulsions by decompression of phosphates on the opposite (convex) face can be accommodated without the normal cost of phosphate crowding upon collapse of the concave face. We have been interested in extending this reasoning to a double-helical model of DNA. In particular, we wished to exclude the possibility that there is actually no unfavorable electrostatic contribution to the bending of double-helical DNA. Such would be the case if the unfavorable electrostatic free energy change due to phosphate crowding on the inner face of the DNA bend were fully compensated by the favorable electrostatic free energy change associated with dilution of phosphates upon stretching along the convex face.

To explore phosphate crowding, we compared the spacings of phosphates in linear DNA to a corresponding bent DNA segment derived from the nucleosome crystal structure (*36*). To examine only phosphate electrostatics, special models were created in which only the phosphate groups were charged, and electrostatic interaction energies were calculated according to:

$$E = k\frac{q_1 q_2}{\varepsilon r} \tag{1}$$

where E is the electrostatic energy, q_1 and q_2 are charges separated by distance r, ε is the solvent dielectric constant, and k is the appropriate constant to yield energies in units of kcal/mol. Dielectric constants of 1.0, 4.0 and 78.39 were tested to mimic *vacuo*, protein and water environments, respectively. The goal of this approach is *not* to generate a biologically-relevant estimate of free energy change associated with DNA bending, but to determine i) if there is a penalty in electrostatic interaction energy for bending double-helical DNA, ii) if such a penalty is due to phosphate crowding, and iii) whether this penalty arises from short-range phosphate crowding, long-range phosphate crowding, or both.

To study the first two questions, we calculated and compared the electrostatic interaction energies of the two conformations of the 71-bp DNA segment shown in Fig. 1. Calculations were performed without a distance cutoff to compare the interaction energy of equal numbers of phosphates in the bent vs. linear DNAs. The use of distance cutoffs less than the length (240 Å) of the linear DNA will undesirably result in a comparison of the interaction energy of more mutually interacting phosphates in the bent DNA than in the linear DNA. As shown in Table 1, the electrostatic interaction energy of the bent DNA with atomic charges assigned to *all* atoms of the DNA is 1.14-fold higher than that of the corresponding linear DNA regardless what dielectric constant is used. These results suggest that there is a penalty in electrostatic interaction energy for bending the DNA double helix in a homogeneous environment. It is worth noting that this electrostatic penalty would differ when DNA bends in a heterogeneous environment such as over a globular protein, where inner phosphates may experience a less polar environment than outer phosphates (*35*). A more sophisticated calculation would be required to address such a heterogeneous environment using different dielectric constants or an explicit solvent model and protein structure [e.g. (*35*)]. Interestingly, the electrostatic interaction energy of the bent DNA with atomic charges assigned *only to phosphate diesters* is 1.1-fold higher than that of the corresponding linear DNA regardless what dielectric constant is used (Table 1). These results suggest that when charges are assigned to all atoms, the electrostatic penalty for DNA bending is mainly due to phosphate crowding, and that the polarization of the bent DNA contributes less than 30% of the penalty. That this result is independent of the solvent dielectric constant follows naturally from Eq. 1, as the dielectric constant cancels when the ratio, E_B/E_L, is computed.

Although the computed electrostatic energy data in Table 1 are expressed in kcal/mol, these values are difficult to interpret physically and should not be extrapolated quantitatively to aqueous solutions of DNA and ions. Rather, the calculations provide a qualitative sense of geometrical aspects of phosphate crowding on the inner and outer faces of curved DNA.

To resolve whether the electrostatic energy penalty for DNA bending arises from short- or long-range phosphate crowding, we selected 35 inner and 35 outer phosphates in the bent DNA (Fig. 1, right). The electrostatic energetics of the phosphates were compared using *in vacuo* calculations because, as demonstrated above, the fractional electrostatic penalty remains the same in

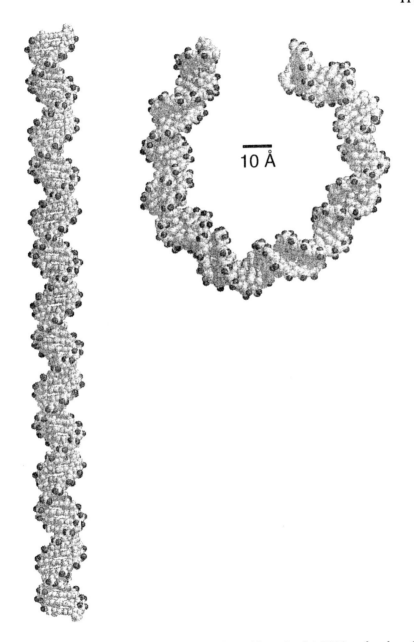

10 Å

Figure 1. Molecular models of linear (left) and bent (right) DNA molecules of identical sequence, 71-bp in length. Anionic phosphate oxygens are shown in black. The DNA sequence and bent geometry (right) were extracted from the X-ray crystal structure of the nucleosome [1aoi.pdb (36)].

Table 1. Electrostatic Interaction Energy of Phosphate Crowding as a Function of DNA Shape[1]

Charged Atoms	Electrostatic Energy[2] (kcal / mol)								
	$\varepsilon=1$			$\varepsilon=4$			$\varepsilon=78.39$		
	linear (E_L)	bent (E_B)	E_B/E_L	linear (E_L)	bent (E_B)	E_B/E_L	linear (E_L)	bent (E_B)	E_B/E_L
PO_3	58018	64012	1.10	14505	16003	1.10	740	817	1.10
All	65297	74627	1.14	16324	18657	1.14	833	952	1.14

1. One supercoil of bent DNA (Fig. 1) was extracted from the nucleosome crystal structure (1aoi.pdb). Electrostatic interaction energy calculations were performed without a distance cutoff as described in Materials and Methods.

2. Energy (E) was calculated for the indicated dielectric constants. Subscripts L and B refer to linear and bent DNA conformations, respectively (Fig. 1).

different homogeneous environments. To ensure the comparison was made with the same number of phosphates in the inner and outer regions of the bent DNA, we extended the calculated electrostatic interaction energies for the two sets of phosphates to long distance cutoffs of 100 Å and 200 Å and obtained a converged electrostatic interaction energy ratio (E_{inner} / E_{outer}) of 1.20 (Table 2). This ratio is comparable to the ratio (1.14) obtained previously in comparing curved and linear DNA conformations. As shown in Table 2, a calculation with a very short distance cutoff of 4 Å yielded an electrostatic interaction energy ratio of 1.0, proving that an equal number of phosphates were included in each set. With the increase of the cutoff from 4 Å to 20 Å to measure interphosphate repulsions, the electrostatic interaction energy ratio (E_{inner} / E_{outer}) increases to a maximum of 1.29 at a cutoff of 7.0 Å and then polytonally declines to the converged ratio of 1.20 (Table 2). These results suggest that the electrostatic penalty for DNA bending is primarily due to short-range phosphate crowding arising from phosphates separated by 4-17 Å. The long-range phosphate crowding is insignificant because the electrostatic ratio at a cutoff of 17 Å (a distance less than the 50 Å radius of the outer phosphates in the bent DNA) equals the electrostatic ratio at 200 Å cutoff (Table 2).

In summary, the simple calculations presented here suggest that there is indeed a penalty in electrostatic interaction energy for bending the DNA double helix. This penalty for bending is due to crowding of phosphates separated over the range of 4-17 Å. We conclude that the electrostatic cost of phosphate crowding on the inner (concave) surface of bent DNA is incompletely compensated by phosphate dilution on the outer (convex) DNA surface.

Experiment: DNA bending upon asymmetric enhancement of phosphate repulsion

We were intrigued by the theoretical prediction that local asymmetric reduction of the solvent dielectric constant in the DNA major groove could enhance phosphate-phosphate repulsions across the groove and alter DNA structure (35). Some related experimental support for this proposition has come from our previous studies that modified the charge of amino acids positioned near the DNA major groove by the binding of bZIP proteins to DNA (25-27). These experiments involve charge variants of bZIP proteins. All of these proteins presumably position alpha-helices in the DNA major groove, displacing water. Our recent focus has not been on local dielectric changes, but on apparent DNA bending observed in response to the charge of three amino acids just beyond the DNA major groove. Our experiments have measured DNA bending away from anionic amino acid side chains positioned near the major groove (25-27). One interpretation of these results is that increased interphosphate repulsion accompanies placement of additional anions in the groove (Fig. 2A).

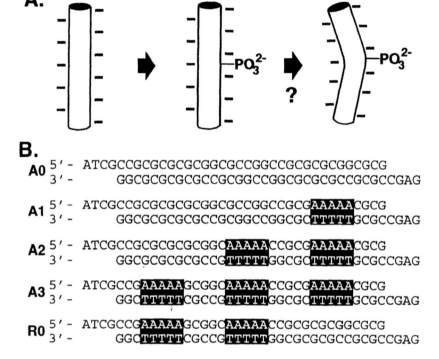

B.

A0　5' - ATCGCCGCGCGCGCGGCGCCGGCCGCGCGCGGCGCG
　　3' - 　　　GGCGCGCGCGCCGCGGCCGGCGCGCGCCGCGCCGAG

A1　5' - ATCGCCGCGCGCGCGGCGCCGGCCGCG AAAAA CGCG
　　3' - 　　　GGCGCGCGCGCCGCGGCCGGCGC TTTTT GCGCCGAG

A2　5' - ATCGCCGCGCGCGCGGC AAAAA CCGCG AAAAA CGCG
　　3' - 　　　GGCGCGCGCGCC TTTTT GGCGC TTTTT GCGCCGAG

A3　5' - ATCGCCG AAAAA GCGGC AAAAA CCGCG AAAAA CGCG
　　3' - 　　　GGC TTTTT CGCCG TTTTT GGCGC TTTTT GCGCCGAG

R0　5' - ATCGCCG AAAAA GCGGC AAAAA CCGCGCGCGGCGCG
　　3' - 　　　GGC TTTTT CGCCG TTTTT GGCGCGCGCCGCGCCGAG

Figure 2. Experimental design. A. Hypothetical DNA bending by electrostatic collapse away from a locus of higher anion density. In this model, an added dianionic phosphonate residue is tethered in the major groove on one face of the DNA double helix. B. Five synthetic DNA duplex standards for calibration of a semi-synthetic DNA phasing assay for monitoring DNA bending (40). C. DNA duplex containing the indicated major groove modifications at position "R". D. Molecular models depicting the tethered phosphonate residue (arrow) in duplex R2. A₅ tracts are shaded. In the upper model, A₅ tracts would narrow the minor grooves on the upper face of the molecules, while the tethered phosphonate dianion is hypothesized to widen the major groove on the bottom face of the molecule. The lower model is rotated 90° about the helix axis so that the tethered phosphonate residue is seen projecting toward the viewer from the floor of the major groove.

Continued on next page.

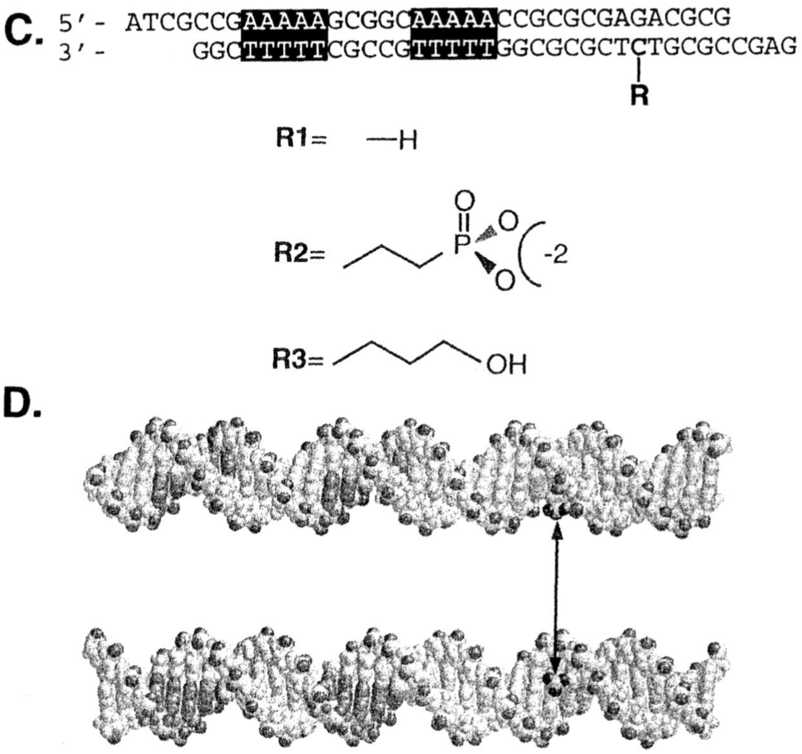

C.
```
5' - ATCGCCGAAAAAGCGGCAAAAACCGCGCGAGACGCG
3' -     GGCTTTTTCGCCGTTTTTGGCGCGCTCTGCGCCGAG
                                        R
```

R1=   —H

R2= (propyl group)—$\overset{\overset{\text{O}}{\|}}{\underset{\text{O}}{\text{P}}}\text{O}\,(-2)$

R3= (propyl group)—OH

D.

Figure 2. *Continued.*

Table 2. Electrostatic Interaction Energy of "Inner" vs. "Outer" Phosphates in Bent DNA[1]

Cutoff (Å)	Electrostatic interaction energy of phosphates (kcal/mol)		
	Inner (E_I)	Outer (E_O)	E_I/E_O
4	2230.0	2236.8	1.00
5	2497.6	2313.7	1.08
6	3380.2	2649.3	1.28
7	4438.7	3438.1	1.29
8	4748.0	3753.9	1.26
9	5132.4	4102.5	1.25
10	5857.8	4628.2	1.27
15	7147.2	5888.8	1.21
16	7299.3	6038.8	1.21
17	7459.3	6205.1	1.20
20	8166.2	6781.4	1.20
100	18123.1	15134.4	1.20
200	18123.1	15134.4	1.20

1. One supercoil of bent DNA (Fig. 1) was extracted from the nucleosome crystal structure (1aoi.pdb). 35 "inner" phosphates and 35 "outer" phosphates were manually selected. Electrostatic energy calculations were performed with a dielectric constant of 1.0 and the indicated distance cutoff as described in Materials and Methods.

To address more specifically whether enhanced interphosphate repulsions in the DNA major groove can cause DNA bending away from that site, we designed a chemical modification that would asymmetrically increase the negative charge density on one DNA surface, and then implemented an electrophoretic phasing assay to measure the extent and direction of any resulting axial bending. These studies were performed in the context of one or more phased A_5-tracts, known to induce ~18° of DNA curvature per A_5-tract toward the minor groove when it is viewed in a reference frame 0.5 bp 3' from the center of the A_5-tract [(41), Fig. 2B]. This approach is reminiscent of our previous studies of DNA bending by the tethering of multiple cationic functional groups across a DNA minor groove (18-20). Two modifications were made for the present experiments.

First, rather than covalent addition of one or more cations, we developed an approach to phosphonylate the exocyclic amino group of a specific cytosine such that two negative charges were added site-specifically to the floor of the DNA major groove (Fig. 2C,D). A monophosphonate adduct was selected because its pK_a values provides two negative charges near neutral pH. For this purpose, we adopted a strategy for post-synthetic pyrimidine modification (42), in which the deoxyuridine 4-position is modified by a triazolyl group (Fig. 3A). This substitution provides an intermediate of sufficient stability to persist during oligonucleotide synthesis, but then serves as an excellent leaving group in an S_N2 reaction with a primary amine derivative of the desired charged groups (or controls) while the synthetic oligomer is still attached to the solid support (Fig. 3A). After work-up, the oligomer can be hybridized to its complement such that the canonical Watson-Crick C-G bp (Fig. 3B) is modified (Fig. 3C). Derivatives bearing neutral or dianionic functions on tethers (Fig. 2B) were prepared in this manner.

Second, we implemented a semi-synthetic electrophoretic phasing assay (40) to monitor the magnitude and direction of any induced DNA bending. This electrophoretic method is highly sensitive and requires many fewer synthetic oligonucleotides than our previous DNA shape studies employing ligation ladders (12, 41, 43). The semi-synthetic DNA phasing assay involves annealing a single synthetic DNA duplex containing modified or unmodified strands, with specific terminal nucleotide overhangs for selective enzymatic ligation to a set of much longer restriction fragments, producing a family of ~250-bp probes. Each probe positions the test duplex at a different distance from a defined element of DNA curvature caused by a phased array of A_5-tracts. As this distance is systematically varied, the resulting differences in probe shape result in different probe mobilities through native polyacrylamide gels. When plotted, these data can be fit to a function whose amplitude is directly related to the curvature of the test duplex. The direction of curvature can be extracted from the analysis. This assay system is ideal for these experiments because i) only a single DNA duplex is required for each modification to be tested (compared to the requirement for up to 14 oligonucleotide pairs for a conventional ligation ladder experiment), ii) the power of phasing analysis can be applied to cases where the test DNA cannot be molecularly cloned, and iii) the assay has been well-calibrated with a number of simple A_5-tract standards (*40*).

The electrophoretic phasing data are shown in Fig. 4. Panel A demonstrates the sensitivity of the assay to DNA shape by comparing mobilities of phasing probes assembled by ligation of arms to the indicated synthetic duplexes containing different numbers of phased A_5-tracts (Fig. 2B). As the degree of test curvature increases from essentially zero (Fig. 4A, duplex **A0**, lanes 1-5) to 54° (Fig. 4A, duplex **A3**, lanes 16-20) phase-dependent mobility changes increase dramatically. This behavior of A_5-tract standards is shown in Fig. 4C, where the mobility of each complex is normalized to the average mobility of the group of five phasing constructs, plotted against the bp spacing in each construct, and fit to an oscillating phasing function. The inset to Fig. 4C illustrates that normalized phasing amplitude is proportional to the predicted DNA curvature under these conditions. An equation fit through this inset data provides a direct approach to estimating DNA curvature for constructs containing tethered anions or controls in place of one of the A_5-tracts.

Phasing data for duplexes **R0** (unmodified general sequence), **R1** (unmodified test sequence), **R2** (tethered phosphonate dianion in test sequence), and **R3** (tethered neutral propanol in test sequence) are shown in Fig. 4B, analyzed in Fig. 4D, and quantitated in Table 3. These results demonstrate that the appended phosphonate function in duplex **R2** results in a 14.3° bend in the DNA. The apparent bend direction is away from the modification, as anticipated, consistent with major groove widening at the site of the added groove dianion.

We synthesized and studied a control duplex in an effort to estimate what fraction of the DNA bending induced by phosphonate tethering might be electrostatic in origin. Analysis of the neutral propanol derivative in duplex **R3** provided interesting insight (Table 3). This neutral analog also caused detectable DNA bending away from the modification, but by an average of only 9.6°. One interpretation of this result is that the average differential bending due to the dianionic function in **R2** vs. the neutral function in **R3** (4.7° net) reflects enhanced electrostatic repulsion in the major groove of **R2**.

To our knowledge, explicit predictions of the degree of DNA bending expected in response to deliberate anion crowding have not been calculated. For the opposite case, namely divalent cation residence in the DNA major groove, Rouzina and Bloomfield (*28*) predict strong (20° - 40°) bending toward the transiently-bound ion, spread over ~6 bp. This model invokes transient repulsion of condensed cations from the vicinity of the bound divalent ion, and strong collapse of the unshielded phosphates upon the divalent ion. It is interesting to compare this model with the consequences of holding a divalent anion in the DNA groove, as we have done. In our experimental model, the local increase in negative charge density might be expected to enhance local

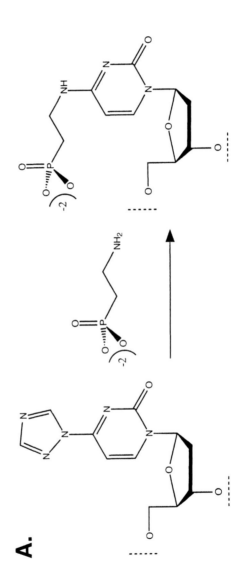

A.

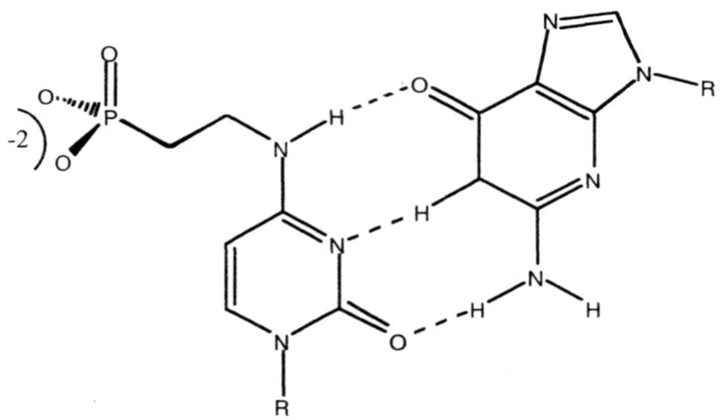

Figure 3. DNA modification chemistry. A. Post-synthetic tethering of dianionic phosphonate to the exocyclic amino group at the 4 position of cytosine by conversion of 4-triazolyl-dU. B. Canonical Watson-Crick C-G base pair. C. Modified C-G base pair with tethered phosphonate residue projecting into the major groove.

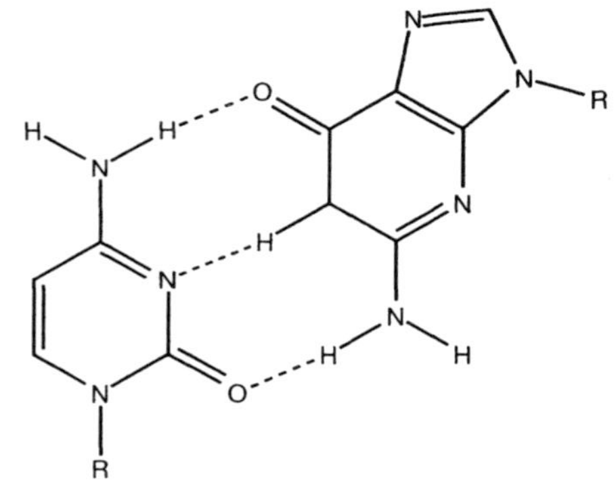

Table 3. DNA Bending by a Tethered Phosphonate Dianion
in the Major Groove

	R0 (unmodified)	R1 (unmodified)	R2 (tethered phosphonate)	R3 (neutral tether)
DNA bend angle (°)[1]	31.0 ± 0.9	33.9 ± 0.1	48.2 ± 0.7	43.5 ± 0.7
bend angle (°) relative to R0[2]	-	0.0 ± 0.1	14.3 ± 0.7	9.6 ± 0.7
bend direction[3] (bp)	16.3	16.2	16.1	16.2

1. DNA bend angle estimated from curve fitting of electrophoretic phasing data and comparison with A_5-tract standards of known curvature (see Materials and Methods). Data reflect the average ± standard deviation based on two measurements

2. Obtained by subtraction of the **R0** bend angle and error propagation.

3. Defined as the distance in bp between the center of curvature of the 5' A_5 tract in the right phasing arm and the 3' terminus of the duplex region of the oligonucleotide insert that minimizes probe mobility (hence maximizing alignment of the induced bend with A_5 tracts). Values of bend direction near 16.2 bp place all A_5 tracts in cis, suggesting that bending induced by phosphonate addition widens the major groove by roll in the expected direction.

128

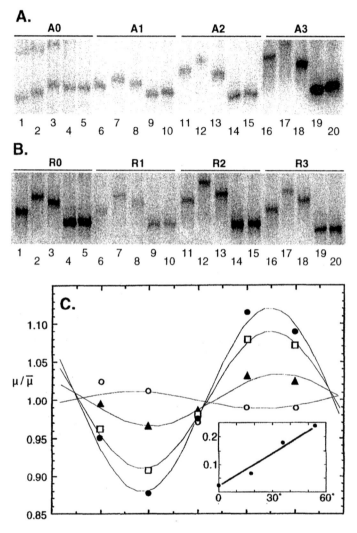

Figure 4. DNA bending by tethered anions. A. Semi-synthetic electrophoretic phasing data for A5-tract standards A0 - A3. B. Semi-synthetic electrophoretic phasing data for DNA duplexes R0 - R3. C. Quantitation and fitting of phasing data from panel A: A0 (open circles), A1 (filled triangles), A2 (open squares), A3 (filled circles). Inset depicts linear relationship between phasing amplitude (vertical axis) and predicted A5-tract curvature (horizontal axis). D. Quantitation and fitting of phasing data from panel B: R0 (open circles), R1 (filled triangles), R2 (filled circles), R3 (open squares).

Continued on next page.

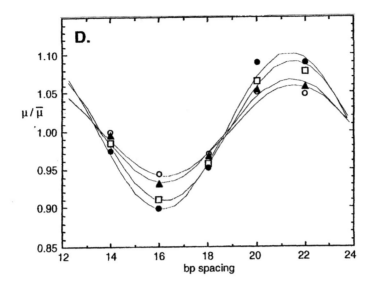

Figure 4. *Continued.*

cation condensation in the vicinity of the tethered divalent anion. This effect would tend to dampen interphosphate repulsions by increased cation screening. Perhaps this effect moderates the extent of DNA bending that might be expected due to tethered anions.

Our observation that perhaps one third of the DNA axial deflection away from a tethered phosphonate dianion is electrostatic in origin presumably represents an upper limit in this case. For example, we cannot rigorously exclude that the 4.7° average difference between the shapes of DNAs modified by dianionic vs. neutral functions is due to subtle steric effects unrelated to electrostatics. As in all chemical analog studies, it is impossible to create neutral control isosteres that mimic all aspects of a charged target compound. Extending this work to other charged analogs may clarify this result.

It is interesting to note that the observed 9.6° of bending away from the neutral propanol derivative attached to the cytosine exocyclic N4 position suggests that modestly bulky ligands at this position may alter normal base pairing and stacking. Such perturbations were generally not observed for uncharged pendant groups at the 5-position of pyrimidines (*18-20*). Experimental designs involving addition of anions to the pyrimidine 5-position may provide an interesting comparison with the present work. Such studies could help to delineate with greater certainty the role of electrostatics in DNA bending away from anions held in the major groove.

Acknowledgements

We acknowledge the synthetic expertise of A. Lebedev, as well as the services of The Mayo Molecular Biology and Proteomics Core Facilities, particularly M. Doerge and L. Benson. We thank L. Cassiday and R. McDonald for comments on the manuscript, and G. Manning, D. Beveridge, R. Lavery, N. Hud, D. Wilson, B. Gold, L. McFail-Isom and L. Williams for encouragement and stimulating discussions.

References

1. Manning, G. S. *Quart. Rev. Biophys.* **1978**,*2*, 179-246.
2. Manning, G. S.;Ray, J. *J. Biomol. Struct. Dyn.* **1998**,*16*, 461-76.
3. Manning, G. S. *Biophys. Chem.* **2002**,*101-102*, 461-73.
4. Record, M. T., Jr.; Zhang, W. T.;Anderson, C. F. *Adv. Protein Chem.* **1998**,*51*, 281-353.
5. Record, M. T.; Anderson, C. F.;Lohman, T. *Q. Rev. Biophys.* **1978**,*11*, 103-178.
6. Saecker, R. M.;Record, M. T., Jr. *Curr. Opin. Struct. Biol.* **2002**,*12*, 311-319.
7. Paulson, M. D.; Anderson, C. F.;Record, M. T., Jr. *Biopolymers* **1988**,*27*, 1249-1265.
8. Maher, L. J. *Curr. Opin. Chem. Biol.* **1998**,*2*, 688-94.
9. Williams, L. D.;Maher, L. J. *Annu. Rev. Biophys. Biomol. Struct.* **2000**,*29*, 497-521.
10. Mirzabekov, A. D.;Rich, A. *Proc. Natl. Acad. Sci. USA* **1979**,*76*, 1118-1121.
11. Manning, G. S.; Ebralidse, K. K.; Mirzabekov, A. D.;Rich, A. *J. Biomolec. Struct. Dyn.* **1989**,*6*, 877-889.
12. Strauss, J. K.;Maher, L. J. *Science* **1994**,*266*, 1829-1834.
13. Strauss-Soukup, J. K.;Maher, L. J. *Biochemistry* **1997**,*36*, 8692-8698.
14. Strauss-Soukup, J. K.;Maher, L. J. *J. Biol. Chem.* **1997**,*272*, 31570-31575.
15. Strauss-Soukup, J. K.; Rodrigues, P. D.;Maher, L. J. *Biophys. Chem.* **1998**,*72*, 297-306.
16. Tomky, L. A.; Strauss-Soukup, J. K.;Maher, L. J. *Nucleic Acids Res.* **1998**,*26*, 2298-2305.
17. Hardwidge, P. R.; Zimmerman, J. M.;Maher, L. J., III. *Nucleic Acids Res.* **2002**,*30*, 1879-1885.

18. Strauss, J. K.; Roberts, C.; Nelson, M. G.; Switzer, C.;Maher, L. J. *Proc. Natl. Acad. Sci. USA* **1996**,*93*, 9515-9520.

19. Strauss, J. K.; Prakash, T. P.; Roberts, C.; Switzer, C.;Maher, L. J. *Chem. & Biol.* **1996**,*3*, 671-678.

20. Hardwidge, P. R.; Lee, D. K.; Prakash, T. P.; Iglesias, B.; Den, R. B.; Switzer, C.;Maher, L. J. *Chem. Biol.* **2001**,*8*, 967-80.

21. Kerppola, T. K.;Curran, T. *Mol. Cell. Biol.* **1993**,*13*, 5479-5489.

22. Paolella, D. N.; Palmer, C. R.;Schepartz, A. *Science* **1994**,*264*, 1130-1133.

23. Strauss-Soukup, J. K.;Maher, L. J. *Biochemistry* **1997**,*36*, 10026-10032.

24. Paolella, D. N.; Liu, Y.;Schepartz, A. *Biochemistry* **1997**,*36*, 10033-10038.

25. Strauss-Soukup, J.;Maher, L. J. *Biochemistry* **1998**,*37*, 1060-1066.

26. Hardwidge, P. R.; Wu, J.; Williams, S. L.; Parkhurst, K. M.; Parkhurst, L. J.;Maher, L. J. *Biochemistry* **2002**,*41*, 7732-7742.

27. Hardwidge, P. R.; Kahn, J. D.;Maher, L. J. *Biochemistry* **2002**,*41*, 8277-8288.

28. Rouzina, I.;Bloomfield, A. *Biophys. J.* **1998**,*74*, 3152-3164.

29. Hud, N. V.; Sklenar, V.;Feigon, J. *J. Mol. Biol.* **1999**,*286*, 651-660.

30. Howerton, S. B.; Sines, C. C.; VanDerveer, D.;Williams, L. D. *Biochemistry* **2001**,*40*, 10023-10031.

31. Hamelberg, D.; McFail-Isom, L.; Williams, L. D.;Wilson, W. D. *J. Am. Chem. Soc.* **2000**,*122*, 10513-10520.

32. Hamelberg, D.; Williams, L. D.;Wilson, W. D. *J. Am. Chem. Soc.* **2001**,*123*, 7745-55.

33. McFail-Isom, L.; Sines, C. C.;Williams, L. D. *Curr. Opin. Struct. Biol.* **1999**,*9*, 298-304.

34. Young, M. A.; Jayaram, B.;Beveridge, D. L. *J. Am. Chem. Soc.* **1997**,*119*, 59-69.

35. Elcock, A. H.;McCammon, J. A. *J. Am. Chem. Soc.* **1996**,*118*, 3787-3788.

36. Luger, K.; Mader, A. W.; Richmond, R. K.; Sargent, D. F.;Richmond, T. J. *Nature* **1997**,*389*, 251-260.

37. Pearlman, D. A.; Case, D. A.; Caldwell, J. W.; Ross, W. S.; Cheatham, T. E.; Debolt, S.; Ferguson, D.; Seibel, G.;Kollman, P. A. *Comput. Phys. Commun.* **1995**,*91*, 1-41.

38. Cornell, W. D.; Cieplak, P.; Bayly, C. I.; Gould, I. R.; Merz, K. M.; Ferguson, D. M.; Spellmeyer, D. C.; Foc, T.; Caldwell, J. W.;Kollman, P. A. *J. Am. Chem. Soc.* **1995**,*117*, 5179-5197.

39. Puglisi, J. D.;Tinoco, I. *Meth. Enzymol.* **1989**,*180*, 304-325.

40. Hardwidge, P. R.; Zimmerman, J. M.;Maher, L. J. *Nucleic Acids Res.* **2000**,*28*, e102.

41. Crothers, D.;Drak, J. *Meth. Enzymol.* **1992**,*212*, 46-71.

42. Xu, Y. Z.; Zheng, Q.;Swann, P. F. *J. Org. Chem.* **1992**,*57*, 3839-3845.

43. Ross, E. D.; Den, R. B.; Hardwidge, P. R.;Maher, L. J. *Nucleic Acids Res.* **1999**,*27*, 4135-42.

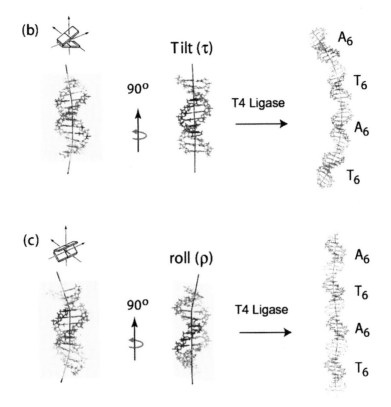

Figure 1. Geometry of Curvature. (b) and (c) are schematic diagrams showing that only when curvature occurs in the direction of tilt at the junctions is the axis of curvature 2-fold symmetric, i.e. rotation of an A-tract by 90° about its helical axis reveals that curvature lies only in one plane. (See text on pages 68-69 for description.)

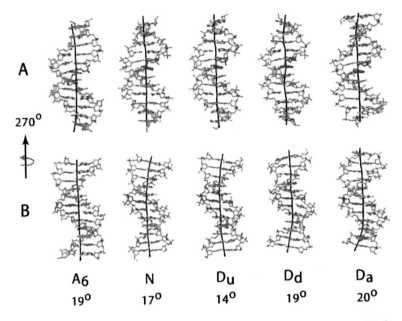

A

270°

B

A₆	N	Dᵤ	D_d	D_a
19°	17°	14°	19°	20°

Figure 3. Comparison of A-tract DNA structures. (See text on page 72 for description.)

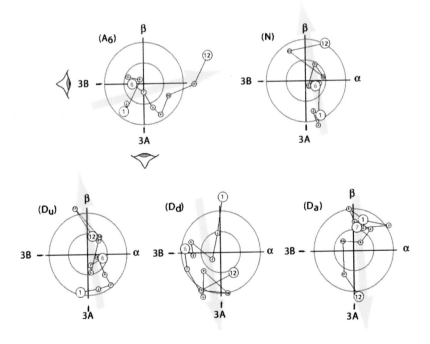

Figure 4. Normal-vector plots of A-tract DNA structures. (See text on page 73 for description.)

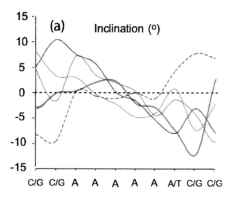

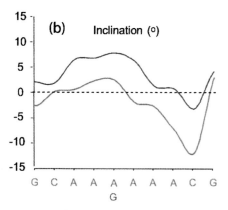

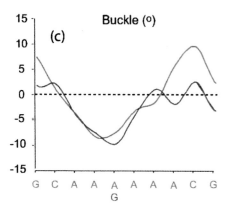

Figure 5. Base inclination within A-tracts. (See text on pages 76-77 for description.)

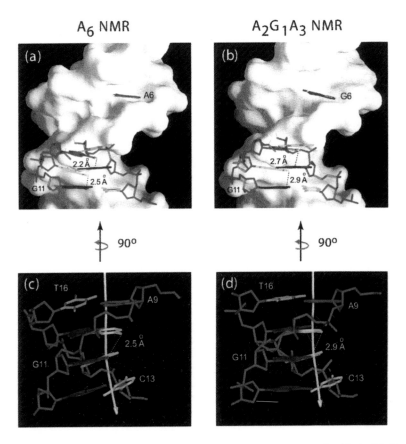

A₆ NMR

(a)

A6

2.2 Å

2.5 Å

G11

A₂G₁A₃ NMR

(b)

G6

2.7 Å

2.9 Å

G11

90°   90°

(c)

T16

A9

2.5 Å

G11

C13

(d)

T16

A9

2.9 Å

G11

C13

Figure 6. (a) and (b) The accessible surface areas of the A₆ and A₂G₁A₃ NMR structures, viewed from the center of each A-tract's minor groove. (c) and (d), H8 of G11 points out roughly perpendicular to the plane of the paper. (See text on page 78 for description.)

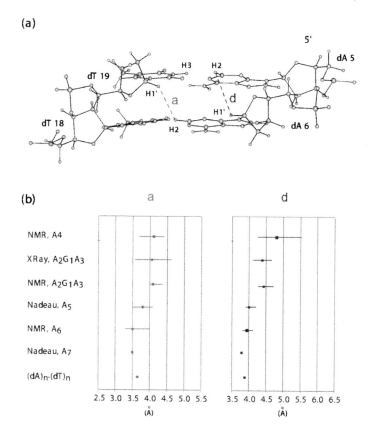

(a)

(b)

Figure 7. Minor groove distances. (See text on page 79 for description.)

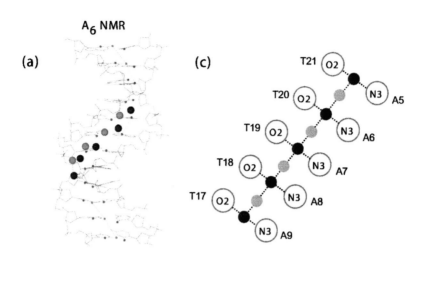

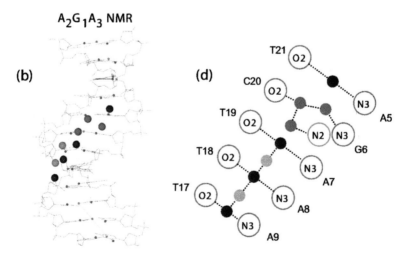

Figure 8. Minor Groove Hydration within A-tracts. (See text on page 81 for description.)

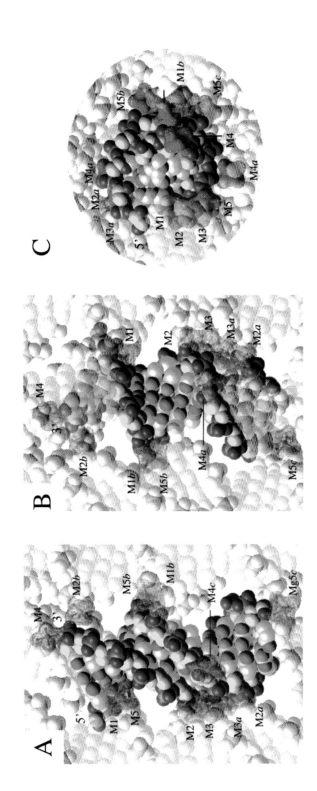

Plate 1. Mg²⁺ environment and packing contacts of the DDD duplex in the Mg form lattice. The dodecamer duplex viewed (**A**) into the minor groove, (**B**) into the major groove and (**C**) roughly along the z-direction of the orthorhombic unit cell. All molecules and metal cations are shown in van der Waals mode. Atoms of the DNA backbone are colored yellow, red and orange for carbon, oxygen and phosphorus, respectively. Nucleobase atoms are colored gray, pink and cyan for carbon, oxygen and nitrogen, respectively. Phosphate groups from neighboring duplexes are highlighted in magenta. Mg²⁺ ions are colored green and are labeled M1 through M5 with small letters in italic font designating individual symmetry mates.

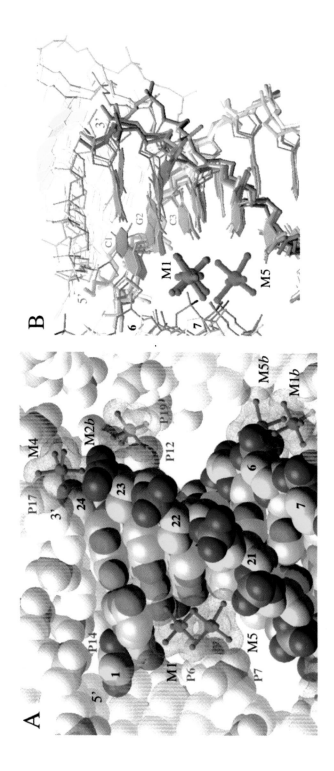

Plate 2. Close-up views of the top half of the DDD duplex (base pairs C1:G24 to A6:T19). (**A**) The duplex viewed across the major and minor grooves, illustrating the kink into the major groove and MgI bound at the G2pC3 (C22pG23) step. The color scheme for atoms is identical to that in plate 1 and Mg^{2+} ions and selected residues as well as phosphate groups from symmetry related duplexes are labeled. (**B**) Superposition of the top halfs of DDD duplexes in selected structures: Nucleic acid database [NDB (40)] code BD0007 [(13), green], BD0005 [(32), cyan], BDL001 [(22), gray] and BDLB04 [(26), pink], demonstrating the similar degrees of kinking into the major groove in the first three structures and the absence of a kink in the structure of the brominated DDD [d(CGCGAATT^{Br}CGCG)]$_2$.

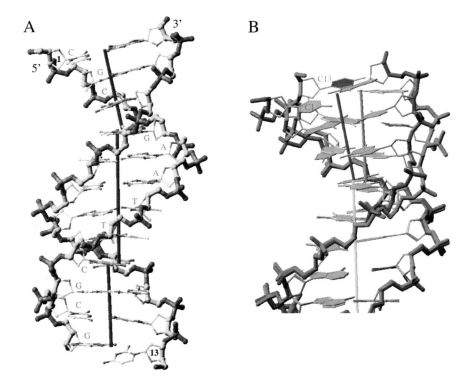

Plate 3. Helical geometry of the DDD duplex in the Mg form [BD0007 (*13*)]. (**A**) The duplex viewed into the minor groove, roughly along the molecular dyad. Helical axes for the CGCG tetramers at both ends and the central hexamer GAATTC were calculated with the program CURVES (*41*) and are depicted in blue. The color scheme for atoms is identical to that in plate 1 and residues of strand 1 are labeled. The drawing illustrates the asymmetric kink of the DDD duplex into the major groove, based in part on a positive roll at the C3pG4 (C21pG22) step. (**B**) Superposition of the top (pink) and bottom (cyan) halves of the DDD duplex. Accordingly, the central hexamer displays almost perfect twofold rotational symmetry, while the kink near one end compresses the major groove and results in deviating geometries of the two 'G-tracts'. The DDD duplex in the orthorhombic lattice can be thought of as composed of a trimer and a nonamer duplex that are themselves straight but exhibit a ca. 11° kink at their interface.

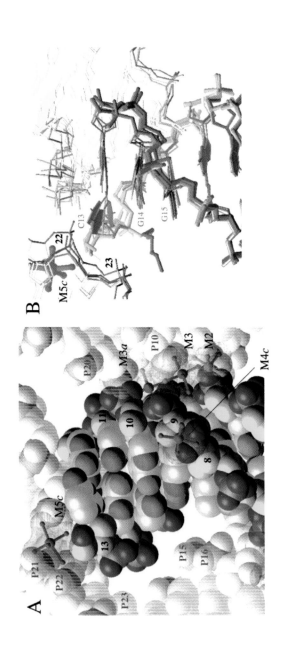

Plate 4. Close-up views of the bottom half of the DDD duplex (base pairs C13:G12 to A18:T7). (**A**) The duplex viewed across the major and minor grooves, illustrating the absence of a kink at the site equivalent to that in the top half of the duplex (plate 2) and the absence of Mg^{2+} in that portion of the major groove. The color scheme for atoms is identical to that in plate 1 and Mg^{2+} ions and selected residues as well as phosphate groups from symmetry related duplexes are labeled. (**B**) Superposition of the bottom halves of DDD duplexes in selected structures: Nucleic acid database [NDB (*40*) code BD0007 [(*13*), green], BD0005 [(*32*), cyan], BDL001 [(*22*), gray] and BDLB04 [(*26*), pink], demonstrating that all four duplexes are straight and that binding of a spermine molecule in one case [only 6 atoms were observed (*32*)] does not affect the helical geometry. The drawing also illustrates the conformational flexibility of the terminal base pair C13:G12, resulting in effective unstacking of C13 in some duplexes. This provides an indication that the constraints due to packing are considerably different for the two DDD duplex ends, the packing around the C13:G12 base pair apparently being less tight.

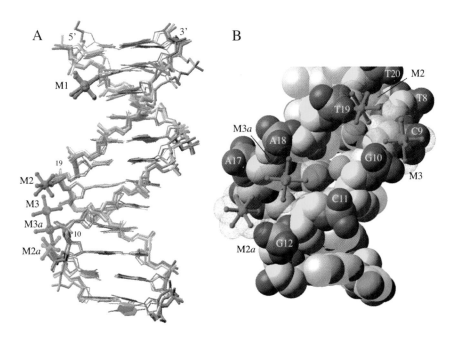

Plate 5. Mg^{2+} coordination in the minor groove of the DDD. (**A**) Superposition of DDD duplexes based on structures BD0007 [(*13*), green], BD0005 [(*32*), cyan], BDL001 [(*22*), gray] and BDLB04 [(*26*), pink], indicating the slight local narrowing of the minor groove at the site of Mg^{2+} coordination in the BD0007 duplex. (**B**) In the structure of the DDD duplex at 1.1 Å resolution, a tandem of Mg^{2+} ions (M2 and M3) crosses the minor groove at the periphery, linking phosphate groups from opposite strands (*13*). Two symmetry-related Mg^{2+} ions (M2a, M3a) get to lie in the vicinity of the minor groove near the G12:C13 end of the same duplex. The color scheme for atoms is identical to that in plate 1 and selected DNA residues are labeled.

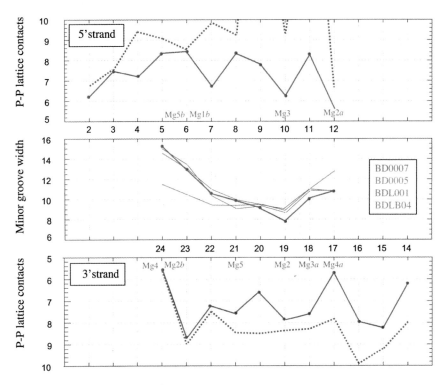

Plate 6. Intra- (defining minor groove width) and inter-duplex P⋯P distances in A in the high-resolution crystal structure of the DDD (13) and Mg^{2+} ions stabilizing closely spaced phosphate pairs in the orthorhombic lattice. *(See text on pages 104 and 105 for description.)*

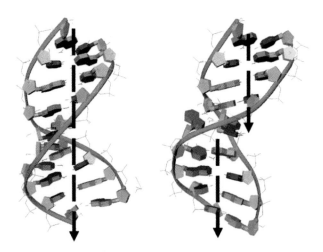

Figure 4: Schematic view of branch site helices in the unmodified (left) and ω-modified (right) duplexes. *(See text on page 170 for description.)*

Chapter 6

Asymmetric Charge Neutralization of DNA

A. Spasic[1], G. Hess[2], K. A. Becker[1], M. Hansen[3], R. Bleil[3], and U. Mohanty[1]

[1]Department of Chemistry, Boston College, Chestnut Hill, MA 02467
[2]Department of Chemistry, Harvard University, Cambridge, MA 02138
[3]University College of Natural Science, Dakota State University, Madison, SD 57042

A quantitative predictive model is formulated that describes the magnitude of bend in DNA oligomers due to asymmetric charge neutralization of the phosphate charges on one side of the DNA double helix. The model, which is a generalization of that due to Manning, takes into account in an approximate way counterion condensation, structural features of the DNA, electrostatic interactions between the charged groups through screened Debye-Hückle potential, and ionic strength of the buffer. As a result of asymmetric neutralization of a fraction of the phosphate groups on one side (histone) of the DNA, the compressive force that act along that side of the DNA is calculated. The in-plane bend angle for two discrete line charges that are in parallel orientation with respect each other are in good agreement with the experimental data of Strauss and Maher for purely monovalent and in mixed solution containing univalent and divalent counterions. Our results for DNA four-way junction suggest the possibility that as a result of asymmetric charge neutralization of DNA, there is twisting, in addition to bending of the molecule.

Introduction

Elucidating the characteristics of bending and flexibility of DNA are of prime importance since it would pave the way towards unraveling the origin of DNA replication and recombination (*1-11*). Understanding the nature of the forces responsible for DNA bending for a given sequence would be invaluable in predicting regulatory protein binding sites.

In a seminal paper that appeared over two decades ago, Mirzabekov and Rich elucidated the importance of electrostatic interactions in the bending of nucleosomal DNA by the eight core histones (12). Mirzabekov and Rich made the conjecture that part of the driving force for bending around nucleosomes is caused by asymmetric charge neutralization of the phosphate groups in nucleosomal DNA (*12,13*). The authors argued that histones neutralized phosphate groups on one side of the helical surface of a nucleosomal DNA, while the free side retained the charges appropriate of the phosphate groups (*12,13*).

The conjecture by Mizabekov and Rich was tested and confirmed in an exquisite series of experiments by Strauss and Maher (*13*). These authors constructed DNA molecules in which selected phosphate groups were replaced by neutral methylphosphonate molecules. Three of these molecules were inserted on each strand so they are located as neighbors across the minor groove (*13*). Consequently, the electrostatic repulsion amongst the phosphate groups is interrupted. The sites neutralized were then phased with respect to A-tracts -- a strategy developed by Crothers and coworkers in their seminal work on sequence directed bends in DNA (*11,13-15*). Comparative gel retardation experiments was then utilized to determine the direction and the magnitude of bending due to asymmetric charge neutralization (*13*). For a 21 bp GC rich segment, an induced bend of $21°$-$25°$ towards the neutral side is observed due to selected neutralization across the minor groove (*13*). The electrostatic nature of the bending was deduced by measuring the induced bend angle as a function of varying concentration of multivalent ions and interpreting the results qualitatively in terms of a counterion condensation model (*16*).

In an elegant and penetrating study by Olson, Manning and coworkers, an all atom energy simulations of DNA dodecamers of alternating d(CG)·d(CG) sequences were carried out in which charges on six of the phosphate groups were either neutralized by neutral methylphosphonate molecules or charge neutralized (*17*). The basic result was that charge neutralization on one side of the DNA double helix results in substantial bending towards the side that is neutralized (*17*). Another significant result is that the average DNA conformation in the presence of methylphosphonate molecules is the same as that produced by charge neutralization (*17*) However, due to lack of salt dependence in the model, quantitative predictions could not be compared with the experimental data of Maher and coworkers (*13*).

In this paper we briefly describe a model that quantitatively predicts the magnitude of bend when asymmetric charge neutralization of the phosphate charges occurs on one side of B-DNA and in Holliday DNA four-way junctions in an aqueous solvent containing monovalent and divalent counterions. The predictions for the bend angles are then compared with experimental data of Strauss and Maher (*13*).

Theoretical Methods

In this section we make use of Manning's counterion condensation model to describe the stretching force acting on the DNA due to polyelectrolyte effects as a result of the asymmetric nature of charge neutralization of the phosphate groups. A generalization of the model for a DNA four-way junction is then briefly described.

Counterion Condensation

Consider a single B-DNA molecule immersed in an aqueous solution of a 1:1 electrolyte at a temperature T. In a seminal work, Manning proposed that the stability properties of the polyelectrolyte are determined by the linear charge density parameter ξ, which is defined to be the ratio of the Bjerrum length l_b to the charge spacing b of the phosphate groups (*16*)

$$\xi = \frac{l_b}{b} \ .$$

(1)

For B-DNA, the linear charge density parameter is 4.2. Since ξ is much larger than unity, Manning argued that the system is inherently unstable. Counterions from the bulk solution would condense on the backbone of the DNA so as to reduce the bare charge of the phosphate groups (*16*).

The counterion condensation model predicts that DNA in aqueous solution is surrounded by counterions that reduce the bare phosphate charges by a factor v. If the two phosphate groups interact via screened Debye-Hückel potential, then the total polyelectrolyte free energy G of the system is a sum of two terms. One of the terms is the electrostatic free energy of the line charge, G_{el}; this is obtained by summing screened Debye-Hückle potential for all pairs of P phosphates, reduced by a factor v due to counterion condensation along the backbone of the DNA. The remaining term is the entropic contribution to G which results from mixing the free cations, the bound cations, and the solvent molecules. This term can be expressed in terms of an effective volume V_p

around the polyion, salt concentration c_s and the fraction of condensed counterions per polyion charge; the latter is obtained by minimizing the total free energy with respect to v, under limiting law conditions (*16*).

Polyelectrolyte Stretching Force

As mentioned in the last section, the B-DNA is represented as a linear array of P univalent charges with equal spacing. Since the phosphate charges repel each other, the polyelectrolyte chain is subjected to a stretching force F defined as $F = -\left(\dfrac{\partial G}{\partial L}\right)$. This force is equal for both DNA strands before any neutralization of the phosphate charges occurs.

Asymmetric Charge Neutralization

To describe asymmetric charge neutralization, the two sides of the DNA are treated separately in calculating the stretching force. This is done because one side of the DNA contains the histone, while the other side is free. Consequently, we denote one side of the helix by h to indicate phosphate neutralized by the histone residue. The other side of the DNA helix is denoted by f — the free side. We denote by F_f and F_h the force on the histone-free side and on the neutralized side due to the histone residues respectively. Let α be the percentage of the phosphate charges that are neutralized on the histone side of the DNA. The neutralized side of the DNA will have contour length L_h, and P_h number of phosphate charges. The distance between the phosphate charges is $b_h(3.4 \text{ Å}/(1-\alpha))$, where h stands for the side with the histone. In contrast, the unchanged side has contour length L_f, P_f number of phosphate charges and the distance between the charges is $b_f(3.4 \text{ Å})$, where the subscript f denotes the free side. Since the contour length of the DNA oligomer is $L=Pb$ for both the histone and free sides, it follows that $P_h = \dfrac{P}{2}(1-\alpha), P_h = \dfrac{P}{2}(1-\alpha)$ and that $P_f = P/2$; further, the spacing between the charges can be expressed in terms of b_h and b_f (*18*)

$$b = \frac{b_h b_f}{b_h + b_f}. \tag{2}$$

When phosphates are neutralized on the histone side, the stretching force on that side is diminished, while the compressive force acts to restore equilibrium by bending the DNA. The forces that act on the histone and free sides can be determined by taking the derivative of the total free energy G with respect to either L_h or L_f and one finds that (18)

$$F_h = F\left(\frac{\partial L}{\partial b}\right)\left(\frac{\partial b}{\partial b_h}\right)\left(\frac{\partial b_h}{\partial L_h}\right) = \left(\frac{1-\alpha}{2-\alpha}\right)F, \qquad (3)$$

$$F_f = F\left(\frac{\partial L}{\partial b}\right)\left(\frac{\partial b}{\partial b_f}\right)\left(\frac{\partial b_f}{\partial L_f}\right) = \left(\frac{1}{2-\alpha}\right)F, \qquad (4)$$

where the corresponding forces when $\alpha = 0$ are denoted as

$$F_f^o = \frac{1}{2}F = F_h^o. \qquad (5)$$

Using Eqs. (3)-(5), the change in force after neutralization that cause the DNA to bend is

$$\Delta F_f = F_f^o - F_f, \qquad (6)$$

$$\Delta F_h = F_h^o - F_h. \qquad (7)$$

Bend Angle

To evaluate the bend angle due to charge neutralization, we utilize a strategy due to Manning that is readily adapted to ionic oligomers (18). The compressive forces ΔF_f and ΔF_h act on opposite sides of the DNA helical axis. The compressive forces are commensurate to a pure compression $(F_h + F_f)$ that is located near the center of the duplex axis, and a bending moment $d(F_h - F_f)$, where d is the distance from the helical axis to the free and the histone sides (18). The pure compression acts in phase with the pure bend to induce, at the center of the oligomer, a radius of curvature τ, that can be expressed in terms of the persistence length, length of the polyion, the distance d of the phosphate group from the duplex axis to the sides f and h and the compressive and bending forces (18)

$$\tau = \frac{B}{d(\Delta F_h - \Delta F_f)} cos\left(\frac{L}{2}\sqrt{\frac{\Delta F_h + \Delta F_f}{B}}\right). \qquad (8)$$

We take a value 9.05 Å for the distance of the phosphate group from the duplex axis (*18*). The Hook's constant B is related to the persistence length λ. As a result of asymmetric charge neutralization of the phosphate charges, the DNA bends by an angle

$$\theta = \frac{180L}{\tau\pi}. \tag{9}$$

Generalization of Manning Model

We have generalized Manning's model to describe asymmetric charge neutralization of the phosphate charges in the presence of purely monovalent and mixed solution to two DNA constructs: (a) two discrete line charges that are in parallel orientation to each other, and (b) two discrete line charges that are oriented at right angles with respect to each other. Case (b) reflects the simplest polyelectrolyte model of a Holliday DNA four-way junction in aqueous solution (*20,21*). DNA four-way type junctions are known to play a fundamental role in homologous genetic recombination (*20*).

Each discrete line consists of P phosphate charges that are interacting by screened Debye-Hückel potential. We then compute the total electrostatic free energy for assembling two line charges, whose centers are at a finite distance and which are oriented with respect to each other at some specified angle, including the entropic contribution from confining the counterions in a volume around the DNA junction. Under suitable conditions, the two line charges will be "bound" at a distance that depends on the charge spacing, the linear charge density, and the ionic strength (*19*). Typically, this distance is of the order of the diameter of B-DNA. By exploiting the Ray-Manning formalism (*19*), we determine the potential of mean force between the two DNA molecules and the counterion condensation characteristics.

Analysis shows that the stretching force in divalent salt solution is approximately given by

$$F = C\{K(4(z(z-1))^2\,\xi(\alpha)-1) + 2z^2(-\ln(2)+\ln(\frac{10^3\xi(\alpha)}{l_b c_s B}) + 2z\xi(\alpha)-1)$$

$$+(4(z(z-1)\xi(\alpha))^2 - 1)\sum_{i=1}^{m-1}\frac{\exp(-i\kappa b(\alpha))}{i}(1-\frac{i}{m})$$

$$+(2z\xi(\alpha))^2\kappa b(\alpha)[1-\frac{1}{z}(1-\frac{1}{2z\xi(\alpha)})]^2\sum_{i=1}^{m-1}\frac{\exp(-i\kappa b(\alpha)}{1}(1-\frac{i}{m})\}$$

(10)

,

where $K=-\ln(\frac{\kappa\rho}{2})-0.5772$, $C=\frac{k_BT}{4z^4\xi(\alpha)b(\alpha)}$, k_B is the Boltzmann

constant, $B=\frac{3}{2}e^{-0.5772}\pi N_A\rho$, T is the absolute temperature, κ is the Debye screening parameter and is related to the ionic strength of the buffer, z is the valence of the counterion, N_A is the Avogadro number, ρ is the distance between the centers of the two DNA molecules, and m is the number of base pairs. The details of the calculation will be presented elsewhere.

Results and Discussion

The structural characteristics of B-DNA used in the model are the number of phosphate charges P, the spacing of the charges b, the Bjerrum length l_b and the persistence length λ. The charge spacing is 1.7 Å and the Bjerrum length is 7.1 Å at room temperature. The distance ρ between the centers of two DNA molecules is taken to be 20 Å for the parallel construct, while the distance d of the phosphate group from the duplex axis is around 9.05 Å. A 21 bp DNA molecule was employed in the experiment by Maher and his coworkers (13). Strauss and Maher used a value 0.14 for the percentage of the phosphate charges that are neutralized on the histone side of the DNA. The TBE buffer consists of 45 mM Tris borate and 1 mM EDTA with pH equal to 8.4. The ionic strength of the buffer deduced from the Henderson-Hasselbach and Davies equation is 0.0256 M. The concentration of added Mg^{2+} cations is 6 mM. The persistence length of B-DNA is around 156bp (22).

For the single line model due to Manning, the bend angle is 6.91° for purely monovalent, while it is 4.59° for mixed solution containing univalent and divalent cations. In contrast, for two line charges that are in parallel orientation with respect to each other, the predicted in-plane bend angles are 15.05° and 7.50° for purely monovalent and for mixed solution, respectively. In contrast, experimental data of Strauss and Mayer yields a value of 21° and 6.7° for purely monovalent and for mixed solution, respectively (13a). We have also investigated how the in-plane bend angle varies with the percentage of the phosphate charges that are neutralized.

We have further examined asymmetric charge neutralization in a DNA four-way junction. As mentioned before, we have modeled a four-way junction as two discrete line charges that are oriented at right angles with respect to each other. Let α be the percentage of the phosphate charges that are neutralized on one of the two discrete line charges. As a result of charge neutralization, the line bends. We assume that the bent line lies in a plane orthogonal to the plane in which the second line charge reside. The predicted bend angles are tabulated in Table 1. The magnitude of the bend angle for this DNA construct suggests that asymmetric charge neutralization of DNA may result in twisting of the molecule, in addition to bending. There are no data on these types of DNA secondary structures and we hope that our results will stimulate further experimental investigations. These results and the details of the calculation will be published elsewhere (23).

Two further generalizations of the model are being examined. In the first approach, we characterize the branched DNA structures as follows. The negative charges of the DNA residues reside on the phosphate atoms. The phosphate atoms are obtained by projection of the helical coordinates onto the symmetry axis of the helix as in the counterion condensation model. However, experiments indicate distortion of DNA helical parameters for several base pairs around the kink sites, while near the four-way and three-way DNA junctions, B-DNA form is usually observed (21). Consequently, the charge spacing of the phosphates near the branch and kink sites will be varied around the canonical value of 1.7 Å, for a distance of several Å from the junction along each arm (21). The geometry of the shapes that are being considered are (a) 4-arm DNA; we are studying square planar, X-shaped, and ideal tetrahedral junctions. In the X-shaped, the two inter-arm angles intersect at 60°. (b) 3-arm DNA; we are also investigating T-shape and Y-shape junctions. The three inter-arm angles intersect at 120° for Y –shape junction, while the two inter-arm angles intersect at 90° for a T-shape junction.

In the second approach, a more accurate model for the electrostatics of the DNA that accounts for the double helical nature of the molecule is being examined. The model is based on the potential of mean force approach ---- an approach that makes use of liquid state physics to describe the interactions of the small ions in solution and the phosphate charges, and yet, is structurally detailed enough to distinguish between the 3' and the 5' ends of a DNA, for example (24). The main advantage of the potential of mean force approach is that it is computationally manageable even with the inclusion of hydration effects and its accuracy is comparable to that of computer simulations of macromolecules, as has been demonstrated in a number of case studies (24). Quantitative results are not yet available for structurally detailed models for comparison with the predictions of the two-DNA constructs studied here.

Table 1. The prediction of the model due to asymmetric charge neutralization of the phosphate charges for two constructs: (a) two DNA molecules with parallel orientation, and (b) two DNA molecules that are oriented at 90° with respect to each other. α is the percentage of the phosphate charges that are neutralized on one side of the DNA.

$\alpha = 0.14$	Bend angle (purely monovalent	Bend angle(mixed solutions
Strauss and Maher [ref.13]	21° (24°-26°)	6.7 ^{o}
Parallel Orientation	15.05 ^{o}	7.50 ^{o}
Holliday four-way junction	34.29 ^{o}	15.03 ^{o}

Conclusion

A quantitative model is proposed that describes the magnitude of folding of DNA oligomers by asymmetric neutralization of the phosphate groups. The model takes into account in an approximate way counterion condensation, structural features of the DNA, ionic effects of the buffer solution, and electrostatic interactions between the charged groups through a screened Debye-Hückle potential. As a result of asymmetric neutralization of the phosphate groups on one side of the DNA helical surface, the compressive force that acts along that side of the DNA is calculated. The predicted in-plane bend angles for two line charges in parallel orientation in monovalent and in mixed solution is in good agreement with the experimental data of Strauss and Maher. The model is generalized to deal with novel DNA conformations, such as a DNA four-way junction. As a result of asymmetric charge neutralization, the importance of accounting for DNA twisting, in addition to bending, is pointed out.

Acknowledgments

The support of the research (U.M.) by the National Science Foundation is gratefully acknowledged. One of us (U.M.) thanks N. Stellwagen for her comments on the manuscript and J. Mayer III for a fruitful discussion during the initial stages of this work and for his encouragement.

References

1. Travers, A.A. *Cell* **1990**, 60, 177.
2. Ripe, K.; Von Hoppel, P. H.; Langowski, J. *Trends Biochemical Science* **1995**, 20, 500.
3. Kahn, J.D.; Crothers, D. M. *Cold Spring Harbor Symposium on Quantitative Biology*, Volume LVIII, Cold Harbor Laboratory Press, **1993**.
4. Hwang, D. S.; Kornberg, A. J. *Journal of Biological Chemistry* **1992**, 267, 23083.
5. Nudler, E.; Avetissova, E.; Markovstov, V.; Goldfarb, A. *Science* **1996**, 273, 211-217.
6. Gartenberg, M. R.; Crothers, D. M. *Nature* (London) **1988**, 333, 824.

7. Crothers, D. M. *Current Biology* **1993**, 3, 675.
8. Goodman, S. D. ; Nicholson, S. C.; Nash, H. A. *Proc. Nat. Acad. Science USA* **1992**, 89, 11910.
9. Zinkel, S. S.; Crothers, D. M. *Journal of Molecular Biology* **1991**, 219, 201.
10. Schultz, S. C.; Shields, G. C.; Steitz, T.A. *Science* **1991**, 253, 1001.
11. Crothers, D. M. *Science* **1994**, 266, 1819-1820.
12. Mirzabekov, A. D.; Rich A. *Proc. Nat. Acad Science USA* **1979**, 76, 1118.
13. (a) Strauss, J. K.; Maher III, J. L. *Science* **1994**, 266, 1829-1834; (b) for a review see Williams, L. D.; Maher, L. J. *Ann. Rev. Biophys. Biomol. Struct.* **2000**, 29, 497-521.
14. Drak, J.; Crothers, D. M. *Proc. Nat.Acad. Science USA* **1991**, 88, 3074-3078
15. Kahn, J. D.; Y. E.; Crothers, D. M. *Nature* **1994**, 308, 163-166.
16. (a) Manning, G. S. *Quarterly Review of Biophysics* **1978**, 111, 179-246. (b) Manning G. S. *Journal of Chemical Physics* **1969**, 51, 924-933.
17. (a) Kosikov,K. M.; Gorin,A. A.; Lu, X. J.; Olson, W.K.; Manning, G. S. *J. Am. Chem. Soc.* **2002**, 124, 4838-4847. (b) see also Sanghani. S.; Zakrzewska, K,; Lavery, R. **1996**. in Sarma, R., and Sarma, M. (eds). Biological Structure and Dynamics. Adenine Press, Schenectady, Vol. 2, pp. 267-278.
18. Manning, G. S.; Ebralidse, K.K.; Mirzabekov, A. D.; Rich, A. J. *Biomolecular Structure and Dynamics* **1989**, 6, 877-889.
19. Ray, J.; Manning, G. S. *Langmuir* **1994**, 10, 2450-2461.
20. Lilley, D. M.; Clegg, R. M. *Quarterly Review of Biophysics* **1993**, 26, 131-175.
21. Fenley, M.; Manning, G.; Olson, W. *Biophys. Chem.* **1998**, 74, 135-152.
22. (a) Manning, G. S. *Cell Biophysics* **1985**, 7, 57-89. (b) Manning, G. S. *Biophysical Chemistry* **2002**, 96, 3963-3969.
23. Spasic, A.; Hess, G.; Becker, K. A.; Boone, M.; Bleil, R.; and Mohanty, U. (unpublished, **2003**).
24. (a) Klement, R.; Soumpasis, D. M.; Jovin, T. M. *Proc. Natl. Acad. Sci. U S A.* **1991**, 88, 4631-5. (b) Klement, R.; Soumpasis, D. M.; Kitzing, E. V.; Jovin, T. M. B*iopolymers* **1990**, 29, 1089-103. (c) Hummer, G.; Soumpasis, D. M. *Phys. Rev. E* **1994**, 50, 5085-5095.

Chapter 7

Structural and Dynamic Properties of Four-Way Helical Junctions in DNA Molecules

David M. J. Lilley

Cancer Research U.K. Nucleic Acid Structure Research Group, Department of Biochemistry, MSI/WTB Complex, The University of Dundee, Dundee DD1 5EH, United Kingdom

Four-way helical junctions in DNA are the central intermediates in genetic recombination. In the presence of divalent metal ions they fold into the stacked X-structure by the pairwise coaxial stacking of helical arms. Exchange between alternative stacking conformers occurs in solution. The structure of four-way junctions is specifically recognised by proteins, notably by the resolving enzymes.

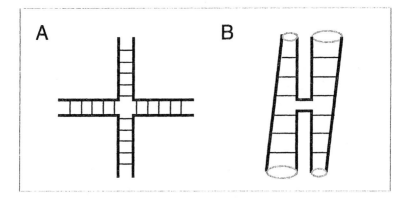

Figure 1. Schematic illustrations of a four-way junction. **A.** The junction is depicted in an open form, showing the connection between four DNA helices. **B.** The junction is shown with coaxial helical stacking, in an antiparallel form with a right-handed rotation. Note that this form has two-fold symmetry, and the strands are divided into continuous and exchanging strands.

Four-way DNA junctions

Four-way DNA junctions created by strand exchange processes (when they are often called Holliday junctions) are the central intermediates of homologous genetic recombination [1-4]. This process is important in the repair of double-strand breaks in DNA, the processing of replication blocks [5, 6], and in the creation of genetic diversity to drive evolution.

They also form the central intermediate of the integrase family of site-specific recombination [7-11]. In the perfect four-way junction (classified as a 4H junction [12]) four DNA helices are connected by the covalent continuity of the component strands (Figure 1). The junctions have presented a number of interesting conformational challenges, and exhibit dynamic properties. In addition, they are substrates for a group of structure-selective proteins.

The stacked X-structure of the four-way DNA junction

The structure of the DNA junction was originally determined in solution by relatively unconventional approaches, before either NMR or crystallography could be applied to the problem. It was found that in the presence of divalent metal ions the DNA junction folds by the pairwise coaxial stacking of arms (see Figure 1B) to generate a two-fold symmetric structure, termed the stacked X-structure (Figure 2). There are two kinds of strands in this structure. The continuous strands pass down the stacked helices, undeviating through the junction to a first approximation. The exchanging strands pass between the stacked helices at the junction. This structure was first proposed on the basis of comparative gel electrophoresis experiments [13], and confirmed soon after by FRET analysis [14, 15]. The pairwise coaxial stacking was consistent with regions of protection against hydroxyl radical attack [16, 17], and the global shape was in agreement with transient electric birefringence experiments [18]. The coaxial stacking was further supported by the demonstration of *Mbo*II cleavage where the recognition and cleavage sites are located in different helical arms [19]. The structure is X-shaped, where the two arms comprise adjacent arms of the junction that are stacked end-on-end at the point of strand exchange through a common axis in each case. The two axes subtend an angle of approximately 60°, and the exchanging strands pass about the smaller angle , ie the structure is antiparallel.

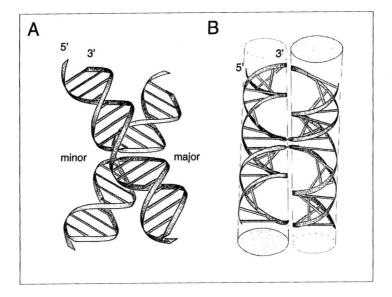

Figure 2. The stacked X-structure of the four-way DNA junction, shown in a schematic ribbon depiction. **A.** The face view of the junction, with the major and minor grooves identified. **B.** View of the major groove side.

148

Molecular modelling [20] suggested that a right-handed cross of helices would generate a favourable juxtaposition between DNA strands and grooves, giving an optimal alignment at 60°. The DNA junction was shown to be right-handed by FRET experiments [14], and the location of the strands in the grooves was in agreement with observed protection against nuclease digestion [21, 22]. The sides of the structure have different structural characteristics. The point of strand exchange has major groove characteristics on one side, and minor groove characteristics on the other. While the early experimental data could not provide information on the local stereochemistry of the junction, molecular modelling showed that that a change in the torsion angle ε about the C3'-O3' bond on the 5' side of the junction from t to g^- would swivel the phosphate group around and redirect the trajectory of the backbone in just the manner required to make the exchange point [20].

In the last three years three crystal structures of DNA junctions (or DNA-

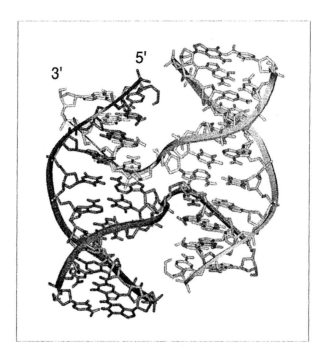

Figure 3. Structure of the four-way DNA junction observed in the crystal [24]. The view is the major groove side of the junction, and the path of the backbones is highlighted by ribbons.

RNA junctions) have been obtained. Curiously all were obtained unexpectedly while trying to crystallise quite different species. Nevertheless, these have provided a wealth of additional conformational data on these branchpoints. All the structures contain right-handed, antiparallel crosses of pairwise, coaxially-stacked helices. The most recent structure from Ho and coworkers is the best to study, since it is devoid of base mismatches (Figure 3). As anticipated by modelling, the trajectory of the backbone required to make the exchange between helical axes is achieved largely by the rotation of a single torsion angle, ie the ε angle (C3'-O3') on the 5' side of the junction changes from t to g^-. This angle has been measured in the crystal to be - 74 [23] and -89° [24]. The geometry of the exchanging backbones orients the central phosphate groups so that the non-bridging oxygen atoms are directed away from the centre with a measured P-P distance is 6.2 Å.

Electrostatics interactions and the role of metal ions

Nucleic acids are highly charged polyelectrolytes, and consequently electrostatic interactions play a rather dominant role in DNA and RNA structure. For this reason metal ions are very important in mediating structural transitions, and this is never more true than in the case of four-way DNA junctions. In the absence of added metal ions, the DNA 4H junction is fully extended, with no coaxial stacking between helical arms. Under these conditions the bases at the point of strand exchange are accessible to attack by chemical agents such as osmium tetroxide [13, 25]. The extended character suggests that metal ions are required to provide charge neutralisation in the stacked X-structure. Thus there is a balance between the stabilisation of the stacked X-structure by the stacking interactions at the point of strand exchange, and the electrostatic interactions tending to open the structure into the extended form. The structure of the low-salt form would be expected to be square, but not necessarily be planar (Figure 4). Since the two sides of the extended structure are inequivalent the structure is probably pyramidal to some degree. The square geometry in the absence of added metal ions was indeed demonstrated by comparative gel electrophoresis [13, 25], and confirmed by FRET experiments [26].

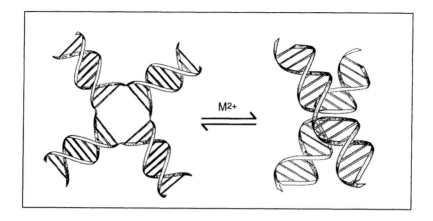

Figure 4. The dependence of the structure of the four-way DNA junction on metal ions. In the absence of added metal ions the junction remains in an extended, square conformation, where the helical arms are unstacked. On addition of divalent metal ions the junction folds into the stacked X-structure.

A variety of ions were shown to bring about the folding [25]. Group II metals such as magnesium and calcium induce folding at concentrations greater than about 100 μM, while complex ions and polyamines are more efficient; 2 μM [Co(NH$_3$)$_6$] (III) or 25 μM spermine are sufficient to promote folding. Group I metal ions, such as sodium or potassium, bring about at least a partial folding of the junction [15], but very high concentrations are required and bases around the centre of the junction remain accessible to addition by osmium tetroxide [25]. The ion-induced folding of the DNA junction can be readily followed by FRET. It was found that the data could be fitted to a two-state model [27], in which folding was induced by the binding of magnesium ions with a $[Mg^{2+}]_{1/2}$ of 7 μM, and a Hill coefficient of $n = 1$. This is consistent with a model in which folding is induced by the non-cooperative binding of one metal ion.

Folding might be induced by site-binding of a specific ion, or diffuse ionic interactions. In the crystal structure of the symmetrical junction [24], there is a box of four phosphates on the minor groove side, comprising the phosphates of the exchanging strands at the point of strand exchange and those immediately to the 3' side. This is clearly a region of high electrostatic potential. We have found that selective neutralisation of these phosphates by substitution by methyl phosphonate can lead to significant perturbation of folding, and neutralisation of both central phosphate groups (P-P separation = 6.2 Å) results in folding into the stacked X-structure in the absence of added metal ions. Hexammine cobalt (III) is particularly efficient in promoting the folding of DNA junctions, and has been footprinted at the junction centre [28]. In the crystal structure of the junction derived from the DNAzyme, a hexammine cobalt (III) ion has been found bound close to the exchanging strands [29]. Van Buuren et al. [30] located a hexammine cobalt (III) ion at an equivalent position in a different four-way junction by means of NOEs between the ammine protons and those of adjacent nucleotides. Electron density consistent with a sodium ion was seen in the crystal structure of the symmetrical junction, positioned close to the oxygen atoms of phosphate groups 5' to the centre, on the major groove side [24].

Dynamic processes in the four-way DNA junction

In principle, the pairwise coaxial stacking of arms in the folding of the four-way junction can occur in one of two ways, that differ in the stacking partners (Figure 5). The strands of the two possible conformers all exchange character, ie the continuous strands in one conformer become the exchanging strands in the other, and vice versa. For a junction of given sequence, the stability of the two conformers would not necessarily be equal, and most junctions studied experimentally exhibit a significant bias to one particular stacking conformer. However, some sequences have been found that do not appear to have a strong bias to one conformer, exemplified by a sequence called junction 7. Using comparative gel electrophoresis we deduced that junction 7 comprised an approximately equimolar distribution of both conformers in rapid exchange [31]. Millar, Chazin and coworkers [32] used a combination of time-resolved FRET and NMR to analyse three DNA junctions, including one that similarly appeared to comprise approximately equal populations of stacking conformers.

152

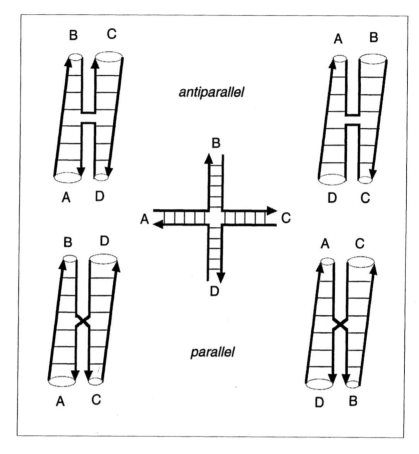

Figure 5. Schematic showing four alternative structures for a given junction. Right and left show alternative stacking conformers, with antiparallel structures at the top and parallel structures at the bottom. All structures are shown right handed, though this feature is arbitrary for the parallel structure. Strand polarity is indicated by arrowheads.

The simultaneous existence of both stacking conformers suggests that interconversion might occur to generate a conformational equilibrium, supported by studies using *Mbo*II cleavage [19, 31]. Altona and colleagues [33] observed interconversion in two DNA junctions using NMR, with exchange rates in the millisecond region.

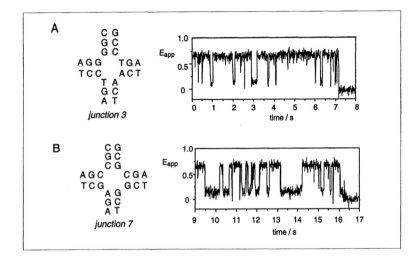

Figure 6. Single-molecule time records for two four-way DNA junctions [34]. The central sequences for junctions 3 and 7 are shown on the left, next to the plots of apparent FRET efficiency recorded over 8-second periods. Note the fluctuations between low- and high-FRET states, corresponding to the two possible stacking conformers. Junction 3 is heavily biased to the conformer corresponding to the high FRET efficiency, while both conformers are populated equally in junction 7.

Direct observation of stacking conformer interconversion is difficult because there is no way to synchronise the process. This is therefore ideally studied in single molecules. We have observed the interconversion process in a strongly biased junction (junction 3), and in junction 7 that has weak bias, using single-molecule FRET spectroscopy [34]. We employed junctions comprising four arms each of 11 bp, in which Cy3 (FRET donor) was attached to the 5'-terminus of one arm, and Cy5 (FRET acceptor) was similarly attached to a different arm. The molecules could be fixed to the surface of a quartz slide via biotin that was attached to a third arm. In one stacking conformer the two fluorophores were widely separated, giving poor FRET efficiency and hence greater donor emission than acceptor. The other conformer resulted in a much closer approach of the two fluorophores, leading to efficient energy transfer resulting in quenching of the donor and sensitised emission from the acceptor. Time traces of apparent FRET efficiency for individual molecules of junctions 3 and 7 are shown in Figure 6. They show clear changes in FRET efficiency with exchange of stacking conformer. This is the first direct observation of the conformer exchange process. Junction 7 displays almost no bias between the two states, while junction 3 exhibits a strong bias towards one stacking conformer, in agreement with previous ensemble studies [13, 31]. The rates of conformer transitions averaged over several hundred transitions from a number of junction molecules were 12 s^{-1} and 3.5 s^{-1} for junction 3 and 5.7 s^{-1} and 6.1 s^{-1} for junction 7 in the presence of 50 mM Mg^{2+} at 25°C. We found that the interconversion rates for junction 7 reduced significantly with increased magnesium ion concentration but backwards and forwards rates remained similar to each other. This indicates that the main effect of changing magnesium ion concentration is in the activation energy, and suggests that the open structure, stable in the absence of Mg^{2+}ions, is a necessary stage in conformer transitions.

Junctions with symmetrical sequences can undergo branch migration, involving the exchange of basepairing partners at the junction and thus moving the point of strand exchange with respect to the sequence. Branch migration over significant distances is an important part of the process of homologous genetic recombination. The kinetics of the process have been extensively studied by Hsieh and coworkers [35-37], who found that the rate was highly dependent on the presence or absence of divalent metal ions. In the presence of magnesium ions step times of 270-300 ms were measured, but in the absence of ions these times reduced to 300 - 400 μs, an acceleration of 1000-fold in the rate of branch migration. This strongly suggests that branch migration must proceed via the open structure of the junction [37], and that stabilisation of the stacked X-structure by divalent metal ions therefore inhibits the process.

It is therefore very likely that the processes of stacking conformer exchange and branch migration pass through the same open structure. Modelling the kinetics of spontaneous branch migration suggests that the process proceeds by a series of steps where a single pair of exchanges is made, consistent with the small step size indicated by the sensitivity to single-base mismatches [38]. Thus branch migration is likely to initiate with an opening into the extended form. One step of basepairing exchange may then occur, randomly in one direction or the other, followed by a return to the stacked X-structure. The transition state for branch migration is likely to involve a greater disruption to the junction than that for conformer exchange, since basepairs must be broken and reformed in the former process, and this is reflected in a greater activation enthalpy for branch migration [36] compared to conformer exchange [34]. The differences in rates for the processes indicate that many conformer transitions should occur before the branchpoint migrates by one step. Thus stacking conformer transitions are an integral feature of branch migration, and a unified energy landscape of these processes has been presented by McKinney et al [34].

Interaction of DNA junctions with proteins

Four-way DNA junctions are specifically recognised by a number of proteins, that exhibit structure-selective binding. In homologous recombination DNA junctions are acted upon by proteins that accelerate and direct branch migration, and by nucleases that achieve resolution. Junction-resolving enzymes have been isolated from a wide variety of sources, from bacteria and their viruses, yeasts, archea and higher eukaryotes and their viruses – these proteins have been recently reviewed [39]. In addition to recognising the structure of the junction, most of these proteins also distort the structure. In the case of RuvA of *Escherichia coli* this can be readily understood in terms of the function, which is to facilitate branch migration. This protein binds as a tetramer, and opens the structure of the junction into an unstacked square geometry that is similar to that present in the absence of metal ions [40, 41]. From the previous section it readily understandable that this would help to accelerate the process of branch migration.

The junction-resolving enzymes all exhibit structure-selective binding, typically binding to junctions in dimeric form with low-nanomolar affinity, not being significantly displaced by 1000-fold higher concentrations of duplex DNA of the same sequence. The basis of this structure-selective recognition is not fully understood, and there are as yet no crystal structures of a complex between any resolving enzyme and a junction.

We have determined the structure of endonuclease I of phage T7 by X-ray crystallography [42]. It comprises two domains separated by an extended β-bridge, each of which contain both polypeptides of the dimer. Interestingly, a result of this is that the active site is formed by amino acids contributed by both polypeptides [43]; these coordinate two metal ions that are involved in the mechanism of phosphodiester bond hydrolysis [44]. On binding the junction endonuclease I distorts both the global and local structure of the DNA, and it has been possible to model the known structure of the protein onto the altered conformation of the junction (Figure 7) [45]. The structure of the protein and the presentation of the active sites have clearly evolved to recognise DNA specifically in the form of the four-way junction. Endonuclease I makes two major cleavages into diametrically opposed strands of the junction, which we deduce to be the continuous strands. When junction 3 (see above) is incubated with endonuclease I, a set of minor cleavages (ie about 20% of total cleavage) are observed on the remaining strands.

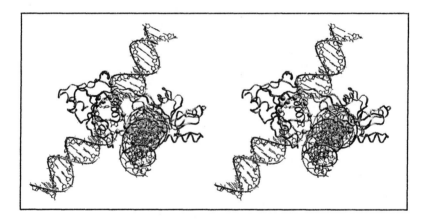

Figure 7. Stereoscopic view of a model of the crystal structure of endonuclease I bound to the distorted structure of the DNA junction [45].

Examination of the single-molecule FRET analysis of this junction (discussed above) shows that about 20% of junction 3 molecules exist in the alternative stacking conformer, and it is therefore clear that the enzyme binds to the junction in whatever conformer it currently adopts, cleaving the continuous strands of that form. It essentially freezes the stacking conformer equilibrium.

It is expected that the recognition processes will turn out to be similar for other junction-resolving enzymes. In each case the protein alters the structure of the junction, yet this seems to be a different distortion in almost every case. The most extreme is that induced by Cce1 of *Saccharomyces cerevisiae*, which opens the global structure into an unstacked square [46] (ie, very similar to that of RuvA above), and also breaks the basepairs that surround the point of strand exchange [47]. The distortion is used by these enzymes in their biological function. In all cases examined these enzymes cleave the two strands sequentially, but with an acceleration for the second cleavage of 10-fold for Cce1 [48] and 150-fold for RuvC [49]. It is likely that this is achieved by release of conformational strain, thereby ensuring a productive resolution of the junction.

Thus in the function of the junction-resolving enzymes, recognition and distortion of the DNA structure are intimately connected to the cleavage of the substrate and the separate aspects cannot be understood in isolation.

How similar are RNA junctions ?

Helical junctions are also important architectural elements in many RNA molecules, and play a particularly important role in autonomously folding small RNA species. For example, junctions are critical to the folding of three of the four known nucleolytic ribozyme; the hammerhead and VS ribozymes contain one and two three-way junctions respectively, while a four-way junction is especially important in the hairpin ribozyme where it lowers the magnesium ion requirement to physiological concentrations [50], and accelerates the folding of the ribozyme by three orders of magnitude [53]. The remaining HDV ribozyme is built around a pseudo-knot structure instead.

Junctions in RNA differ from those in DNA in two main respects. First, the underlying helical structure is different, since RNA helices are A-form whereas DNA tends to adopt the B-form structure. Second, recombination junctions in DNA are generally perfect 4H junctions, but such perfection is not required in an RNA molecule which has usually folded from a single-stranded species. Mismatches and additional bases would be expected to have an important influence on the structure and dynamics of RNA junctions.

However, a number of features do tend to remain common between DNA and RNA junctions. RNA junctions undergo pairwise coaxial stacking of arms to create continuous and exchanging strands similarly to DNA [51, 52]. However, unlike their DNA counterparts, 4H junctions in RNA do not exist in an open form at low salt concentrations, but instead they retain their helical stacking even in the absence of added metal ions. Under these conditions the structure averages to give an approximately perpendicular angle between the axes, but upon addition of divalent metal ions the helices rotate in an antiparallel direction. However, it is clear that RNA junctions are generally more structurally adventurous than the corresponding DNA junctions, and recent single-molecule studies [53] reveal that RNA 4H junction undergo transitions between parallel and antiparallel forms, and between alternative stacking conformers. This dynamic character becomes even more accentuated if unpaired bases are present. The IRES element that allows internal initiation of translation in hepatitis C viral RNA contains a four-way junction with two unpaired bases on one strand, and a single additional base on the diametrically opposed strand (ie a $2HS_2 2HS_1$ junction) [54, 55]. The junction folds by pairwise coaxial stacking of helical arms, into the conformer that locates the unpaired bases on the exchanging strands [56]. However, fluorescence lifetime measurements indicate that in the presence of magnesium ions, the IRES junction exists in a dynamic equilibrium comprising approximately equal populations of antiparallel and parallel species. Curiously, the same junction has recently been crystallised, and the structure solved by X-ray crystallography [57]. In the crystal the junction exhibits the same choice of helical stacking partners, but the structure is almost perfectly parallel. Thus the crystallisation has pulled the parallel-antiparallel equilibrium into the parallel form. Another marked difference between the IRES junction and perfect 4H junctions (including that derived by simply deleting the three additional bases) is the structure in the absence of added metal ions. Under these conditions the IRES junction undergoes a loss of pairwise helical stacking [56], and forms an extended square structure like that adopted by 4H DNA junctions in the absence of added ions.

Acknowledgements

I am grateful to my collaborators and coworkers, whose work I have discussed in this overview, especially Anne-Cécile Déclais, Sophie Liu, Sean McKinney, Alasdair Freeman, Franck Coste, John Hadden, Simon Phillips, Sonya Melcher, Tim Wilson, Elliot Tan and Taekjip Ha. Work in this laboratory is supported by Cancer Research-UK.

References

1. Holliday, R. (1964) A mechanism for gene conversion in fungi. *Genet. Res*, **5**, 282-304.
2. Broker, T. R., and Lehman, I. R. (1971) Branched DNA molecules: intermediates in T4 recombination. *J. Molec. Biol.* **60**, 131-149.
3. Orr-Weaver, T. L., Szostak, J. W., and Rothstein, R. J. (1981) Yeast transformation: a model system for the study of recombination. *Proc. Natl. Acad. Sci. USA* **78**, 6354-6358.
4. Schwacha, A., and Kleckner, N. (1995) Identification of double Holliday junctions as intermediates in meiotic recombination. *Cell* **83**, 783-791.
5. Cox, M. M., Goodman, M. F., Kreuzer, K. N., Sherratt, D. J., Sandler, S. J., and Marians, K. J. (2000) The importance of repairing stalled replication forks. *Nature* **404**, 37-41.
6. Michel, B. (2000) Replication fork arrest and DNA recombination. *Trends Biochem. Sci.* **25**, 173-178.
7. Hoess, R., Wierzbicki, A., and Abremski, K. (1987) Isolation and characterisation of intermediates in site-specific recombination. *Proc. Natl. Acad. Sci. USA* **84**, 6840-6844.
8. Nunes-Düby, S. E., Matsomoto, L., and Landy, A. (1987) Site-specific recombination intermediates trapped with suicide substrates. *Cell* **50**, 779-788.

9. Kitts, P. A., and Nash, H. A. (1987) Homology-dependent interactions in phage λ site-specific recombination. *Nature* **329**, 346-348.

10. Jayaram, M., Crain, K. L., Parsons, R. L., and Harshey, R. M. (1988) Holliday junctions in FLP recombination: Resolution by step-arrest mutants of FLP protein. *Proc. Natl. Acad. Sci. USA* **85**, 7902-7906.

11. McCulloch, R., Coggins, L. W., Colloms, S. D., and Sherratt, D. J. (1994) Xer-mediated site-specific recombination at cer generates Holliday junctions in vivo. *EMBO J.* **13**, 1844-1855.

12. Lilley, D. M. J., Clegg, R. M., Diekmann, S., Seeman, N. C., von Kitzing, E., and Hagerman, P. (1995) Nomenclature Committee of the International Union of Biochemistry: A nomenclature of junctions and branchpoints in nucleic acids. Recommendations 1994. *Eur. J. Biochem.* **230**, 1-2.

13. Duckett, D. R., Murchie, A. I. H., Diekmann, S., von Kitzing, E., Kemper, B., and Lilley, D. M. J. (1988) The structure of the Holliday junction and its resolution. *Cell* **55**, 79-89.

14. Murchie, A. I. H., Clegg, R. M., von Kitzing, E., Duckett, D. R., Diekmann, S., and Lilley, D. M. J. (1989) Fluorescence energy transfer shows that the four-way DNA junction is a right-handed cross of antiparallel molecules. *Nature* **341**, 763-766.

15. Clegg, R. M., Murchie, A. I. H., Zechel, A., Carlberg, C., Diekmann, S., and Lilley, D. M. J. (1992) Fluorescence resonance energy transfer analysis of the structure of the four-way DNA junction. *Biochemistry* **31**, 4846-4856.

16. Chen, J.-H., Churchill, M. E. A., Tullius, T. D., Kallenbach, N. R., and Seeman, N. C. (1988) Construction and analysis of monomobile DNA junctions. *Biochemistry* **27**, 6032-6038.

17. Churchill, M. E., Tullius, T. D., Kallenbach, N. R., and Seeman, N. C. (1988) A Holliday recombination intermediate is twofold symmetric. *Proc. Natl. Acad. Sci. USA* **85**, 4653-4656.

18. Cooper, J. P., and Hagerman, P. J. (1989) Geometry of a branched DNA structure in solution. *Proc. Natl. Acad. Sci. USA* **86**, 7336-7340.

19. Murchie, A. I. H., Portugal, J., and Lilley, D. M. J. (1991) Cleavage of a four-way DNA junction by a restriction enzyme spanning the point of strand exchange. *EMBO J.* **10**, 713-718.

20. von Kitzing, E., Lilley, D. M. J., and Diekmann, S. (1990) The stereochemistry of a four-way DNA junction: a theoretical study. *Nucleic Acids Res.* **18**, 2671-2683.

21. Lu, M., Guo, Q., Seeman, N. C., and Kallenbach, N. R. (1989) DNaseI cleavage of branched DNA molecules. *J. Biol. Chem.* **264**, 20851-20854.

22. Murchie, A. I. H., Carter, W. A., Portugal, J., and Lilley, D. M. J. (1990) The tertiary structure of the four-way DNA junction affords protection against DNaseI cleavage. *Nucleic Acids Res.* **18**, 2599-2606.

23. Ortiz-Lombardía, M., González, A., Erijta, R., Aymamí, J., Azorín, F., and Coll, M. (1999) Crystal structure of a DNA Holliday junction. *Nature Struct. Biol.* **6**, 913-917.

24. Eichman, B. F., Vargason, J. M., Mooers, B. H. M., and Ho, P. S. (2000) The Holliday junction in an inverted repeat DNA sequence: Sequence effects on the structure of four-way junctions. *Proc. Natl. Acad. Sci. USA* **97**, 3971-3976.

25. Duckett, D. R., Murchie, A. I. H., and Lilley, D. M. J. (1990) The role of metal ions in the conformation of the four-way junction. *EMBO J.* **9**, 583-590.

26. Clegg, R. M., Murchie, A. I. H., Zechel, A., and Lilley, D. M. J. (1994) The solution structure of the four-way DNA junction at low salt concentration; a fluorescence resonance energy transfer analysis. *Biophys. J.* **66**, 99-109.

27. Fogg, J. M., Kvaratskhelia, M., White, M. F., and Lilley, D. M. J. (2001) Distortion of DNA junctions imposed by the binding of resolving enzymes: A fluorescence study. *J. Molec. Biol.* **313**, 751-764.

28. Møllegaard, N. E., Murchie, A. I. H., Lilley, D. M. J., and Nielsen, P. E. (1994) Uranyl photoprobing of a four-way DNA junction: Evidence for specific metal ion binding. *EMBO J.* **13**, 1508-1513.

29. Nowakowski, J., Shim, P. J., Prasad, G. S., Stout, C. D., and Joyce, G. F. (1999) Crystal structure of an 82 nucleotide RNA-DNA complex formed by the 10-23 DNA enzyme. *Nature Struct. Biol.* **6**, 151-156.

30. van Buuren, B. N. M., Schleucher, J., and Wijmenga, S. S. (2000) NMR structural studies on a DNA four-way junction: Stacking preference and localisation of the metal-ion binding site. *J. Biomol. Struct. Dynam.* **11**, 237-243.

31. Grainger, R. J., Murchie, A. I. H., and Lilley, D. M. J. (1998) Exchange between stacking conformers in a four-way DNA junction. *Biochemistry* **37**, 23-32.

32. Miick, S. M., Fee, R. S., Millar, D. P., and Chazin, W. J. (1997) Crossover isomer bias is the primary sequence-dependent property of immobilized Holliday junctions. *Proc. Natl. Acad. Sci. USA* **94**, 9080-8084.

33. Overmars, F. J. J., and Altona, C. (1997) NMR study of the exchange rate between two stacked conformer of a model Holliday junction. *J. Molec. Biol.* **273**, 519-524.

34. McKinney, S. A., Déclais, A.-C., Lilley, D. M. J., and Ha, T. (2003) Structural dynamics of individual Holliday junctions. *Nature Struct. Biol.* **10**, 93-97.

35. Panyutin, I. G., and Hsieh, P. (1993) Formation of a single base mismatch impedes spontaneous DNA branch migration. *J. Molec. Biol.* **230**, 413-424.

36. Panyutin, I. G., and Hsieh, P. (1994) The kinetics of spontaneous DNA branch migration. *Proc. Natl. Acad. Sci. USA* **91**, 2021-2025.

37. Panyutin, I. G., Biswas, I., and Hsieh, P. (1995) A pivotal role for the structure of the Holliday junction in DNA branch migration. *EMBO J.* **14**, 1819-1826.

38. Biswas, I., Yamamoto, A., and Hsieh, P. (1998) Branch migration through DNA sequence heterology. *J. Molec. Biol.* **279**, 795-806.

39. Lilley, D. M. J., and White, M. F. (2001) The junction-resolving enzymes.*Nature Rev. Molec. Cell Biol.* **2**, 433-443.

40. Rafferty, J. B., Sedelnikova, S. E., Hargreaves, D., Artymiuk, P. J., Baker, P. J., Sharples, G. J., Mahdi, A. A., Lloyd, R. G., and Rice, D. W. (1996) Crystal structure of DNA recombination protein RuvA and a model for its binding to the Holliday junction .*Science* **274**, 415-421.

41. Roe, S. M., Barlow, T., Brown, T., Oram, M., Keeley, A., Tsaneva, I. R., and Pearl, L. H. (1998) Crystal structure of an octameric RuvA-Holliday junction complex.Molec. *Cell* **2**, 361-372.

42. Hadden, J. M., Convery, M. A., Déclais, A.-C., Lilley, D. M. J., and Phillips, S. E. V. (2001) Crystal structure of the Holliday junction-resolving enzyme T7 endonuclease I at 2.1 Å resolution. *Nature Struct. Biol.* **8**, 62-67.

43. Déclais, A.-C., Hadden, J. M., Phillips, S. E. V., and Lilley, D. M. J. (2001) The active site of the junction-resolving enzyme T7 endonuclease I *.J. Molec. Biol.* **307**, 1145-1158.

44. Hadden, J. M., Déclais, A.-C., Phillips, S. E. V., and Lilley, D. M. J. (2002) Metal ions bound at the active site of the junction-resolving enzyme T7 endonuclease I. *EMBO J.* **21**, 3505-3515.

45. Déclais, A.-C., Fogg, J. M., Freeman, A., Coste, F., Hadden, J. M., Phillips, S. E. V., and Lilley, D. M. J. (2003) The complex between a four-way DNA junction and T7 endonuclease I. *EMBO J.* **22**, 1398-1409.

46. White, M. F., and Lilley, D. M. J. (1997) The resolving enzyme CCE1 of yeast opens the structure of the four-way DNA junction. *J. Molec. Biol.* **266**, 122-134.

47. Déclais, A.-C., and Lilley, D. M. J. (2000) Extensive central disruption of a four-way junction on binding CCE1 resolving enzyme. *J. Molec. Biol.* **296**, 421-433.

48. Fogg, J. M., Schofield, M. J., Déclais, A.-C., and Lilley, D. M. J. (2000) The yeast resolving enzyme CCE1 makes sequential cleavages in DNA junctions within the lifetime of the complex. *Biochemistry* **39**, 4082-4089.

49. Fogg, J. M., and Lilley, D. M. J. (2001) Ensuring productive resolution by the junction-resolving enzyme RuvC: Large enhancement of second-strand cleavage rate. *Biochemistry* **39**, 16125-16134.

50. Zhao, Z.-Y., Wilson, T. J., Maxwell, K., and Lilley, D. M. J. (2000) The folding of the hairpin ribozyme : Dependence on the loops and the junction. *RNA* **6**, 1833-1846.

51. Duckett, D. R., Murchie, A. I. H., and Lilley, D. M. J. (1995) The global folding of four-way helical junctions in RNA, including that in U1 snRNA. *Cell* **83**, 1027-1036.

52. Walter, F., Murchie, A. I. H., Duckett, D. R., and Lilley, D. M. J. (1998) Global structure of four-way RNA junctions studied using fluorescence resonance energy transfer. *RNA* **4**, 719-728.

53. Tan, E., Wilson, T. J., Nahas, M. K., Clegg, R. M., Lilley, D. M. J., and Ha, T. (2003) A four-way junction accelerates hairpin ribozyme folding via a discrete intermediate. *Proc. Natl. Acad. Sci. USA* **100**, 9308-9313.

54. Brown, E. A., Zhang, H., Ping, L. H., and Lemon, S. M. (1992) Secondary structure of the 5' nontranslated regions of hepatitis C virus and pestivirus genomic RNAs. *Nucleic Acids Res.* **20**, 5041-5045.

55. Honda, M., Ping, L. H., Rijnbrand, R. C., Amphlett, E., Clarke, B., Rowlands, D. J., and Lemon, S. M. (1996) Structural requirements for initiation of translation by internal ribosome entry within genome-length hepatitis C virus RNA. *Virology* **222**, 31-42.
56. Melcher, S. E., Wilson , T. J., and Lilley , D. M. J. (2003) The dynamic nature of the four-way junction of the hepatitis C virus IRES. *RNA*, **9**, 809-820.
57. Kieft, J. S., Zhou, K., Grech, A., Jubin, R., and Doudna, J. A. (2002) Crystal structure of an RNA tertiary domain essential to HCV IRES-mediated translation initiation. *Nature Struc. Biol.* **9**, 370-374.

Chapter 8

Role of a Conserved Pseudouridine in Spliceosomal Pre-mRNA Branch Site Conformation

Nancy L. Greenbaum[1,2] and Meredith Newby Lambert[2,3]

[1]Department of Chemistry and Biochemistry, Florida State University, Tallahassee, FL 32306–4390
[2]Institute of Molecular Biophysics, Florida State University, Tallahassee, FL 32306
[3]Current address: Department of Chemistry, University of Michigan, 930 North University, Ann Arbor, MI 48209–1055

We show that the presence of a phylogenetically conserved pseudouridine (ψ)in the pre-mRNA branch site helix of *S. Cerevisiae* induces a dramatically altered architectural landscape compared with that of its unmodified counterpart. The ψ-induced structure places the nucleophile in an accessible position for the first step of splicing. NMR data implicate a water-ψNH1 hydrogen bond in favoring formation of the unique structure of the branch site helix.

The process by which noncoding regions (introns) are spliced from precursor messenger (pre-m)RNA molecules, and the flanking coding regions (exons) ligated, is of fundamental importance in gene expression. Pre-mRNA splicing in eukaryotes is catalyzed by the spliceosome, a ribonucleoprotein machine comprising small nuclear (sn)RNA and protein components[1]. Phylogenetic, mutational, and biochemical studies demonstrate the role of specific RNA sequences in assembly of the catalytic core[2]. Among the critical RNA-RNA interactions is that of a consensus sequence of the intron with a short region of the U2 snRNA. This pairing forms a complementary helix of seven base pairs with a single unpaired adenosine residue. It is the 2'OH of this adenosine, called the branch site, which is the nucleophile in the cleavage reaction at the pre-mRNA 5' splice site. Structural features leading to recognition of the branch site adenosine residue by other components of the catalytic core, and positioning of the 2'OH for nucleophilic activity, are major issues in understanding the structural biology of RNA splicing.

Post-transcriptionally modified bases are a common feature of structural RNAs, such as tRNA, ribosomal RNA, and snRNAs, where they enhance the specificity of RNA-ligand interactions and/or increase thermal stability[3]. Pseudouridine (abbreviated ψ), the most abundant modified base, is formed by enzymatic scission of the nucleoside's glycosidic bond, followed by rotation of the base about its 3-6 axis and reattachment through the carbon at the 5-position of the ring (Figure 1). The modified base, which maintains the same relative position of hydrogen bonding donors and acceptors on its Watson-Crick face as uridine, features an additional ring nitrogen atom that is protonated at physiological pH (NH1). The presence of ψ in defined positions in stems and anticodon loops of tRNA molecules is generally associated with increased stabilization without perturbing structure[4-8]. The increase in thermal stability of RNA secondary structure by ψ has been hypothesized to result from additional hydrogen bonds involving the ψNH1[4,9] or from more favorable stacking interactions between ψ and neighboring bases[8].

ψ residues are prevalent in the 5' segment of U2 snRNA, where some of them have been associated with spliceosome assembly[10]. A ψ residue in the region of U2 snRNA that opposes the 5' neighbor of the branch site adenosine has been identified in all eukaryotes investigated to date[11-14].

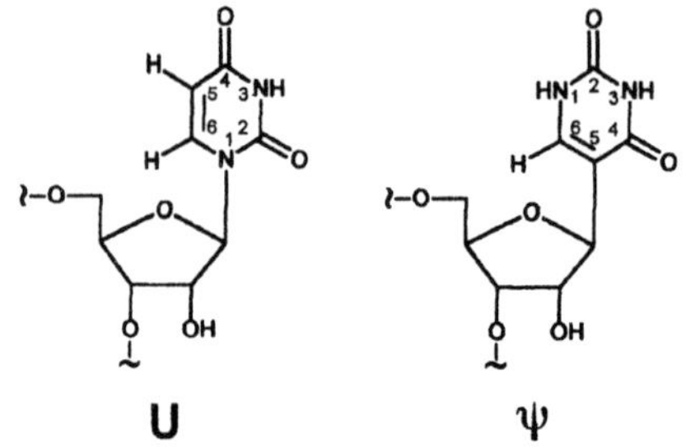

Figure 1. Uridine (U, left) and pseudouridine (ψ, right).

Using NMR spectroscopy, we have determined solution structures of ψ-modified and unmodified duplexes representing the pre-mRNA branch site of *S. Cerevisiae*[15,16]. Structural models reveal that the conserved ψ induces a dramatically different structure than that in the presence of its unmodified counterpart. Our data also support the model of a water-mediated hydrogen bond involving ψ as the source of added stabilization or structural change[17]. The ψ-modified structure is consistent with all previous biochemical and genetic data and may explain how the 2'OH is recognized and positioned for the first step of spliceosomal splicing.

Methods

RNA oligomers for NMR and fluorescence experiments were obtained from Dharmacon, Inc. (Boulder, CO), and were prepared for spectroscopic experiments as previous described[15-17]. Homonuclear NOESY and TOCSY spectra were acquired on a 720 MHz Varian Unity Plus spectrometer (National High Magnetic Field Laboratory, Tallahassee, FL) and a 500 MHz Varian Inova spectrometer (Dept. of Chemistry and Biochemistry, Florida State University), and assigned according to standard methods. Chemical shifts and a number of NOEs supporting structural difference between the two duplexes were reported previously[15]. ROESY-type experiments using a CLEANEX-PM spin-lock pulse train[18] (3kHz spin-lock field) were optimized for application to RNA samples[17]. A torsion angle molecular dynamics (TAMD) protocol[19,20] was utilized in combination with simulated annealing as part of the X-PLOR program[21] for the generation of families of structures of from random starting structures. Coordinates for each structure in the ensemble of 10 lowest energy unmodified structures (1LMV) and the 9 lowest energy structures of the ψ-modified sequence (1LPW) have been deposited in the Protein Data Bank.

Structure of branch site duplexes

The unmodified branch site duplex (uBP) displays a continuous A-type helical geometry, with the branch site adenosine (A24) stacked within the helix. Both A24 and A23, its 5' neighbor, form hydrogen bonds with the opposing U6. The structure (Figure 2) is very similar to that determined for an RNA stem loop representing the phage coat protein-binding site, which differs only in one of the base pairs flanking the extra adenosine residue[22].

A markedly different structure is seen for the ψ-modified duplex (ψBP). The U2 snRNA strand maintains roughly helical parameters, but the backbone of the intron strand in the branch site region has a pronounced kink (Figure 3). The deviation from A-form geometry starts with U22. Chain reversal occurs at the junction between A23 and A24, and the backbone regains helical parameters at C25. H1'-H2' scalar couplings for U22 and A23 suggest non-C3'-*endo* character for these riboses.

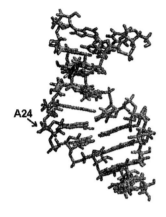

5'$\overset{2}{G}$GUG U$\overset{}{A}$GU$\overset{10}{A}$3'
∘ • • • ∘ ∘ ∘ ∘ ∘
3'$\underset{28}{C}$CAC$\underset{24}{A}$AUCAU$\underset{19}{}$5'   **uBP**

Figure 2: Structural models of uBP, the unmodified branch site duplex. (left) Representative structure (one of an ensemble of 10 low-energy models) of the unmodified branch site duplex. (right) View down the axis of the duplex indicates typical A-type helical geometry. Sequence of uBP is shown on the right: upper and lower strand corresponding to U2 snRNA and the intron sequences, respectively. Bases shown in gray were added to stabilize the helix.

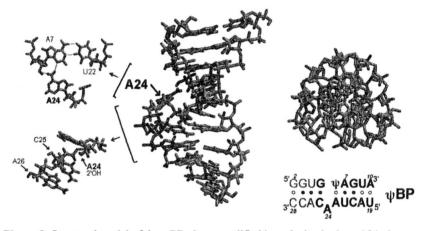

5'$\overset{2}{G}$GUG ψ$\overset{}{A}$GU$\overset{10}{A}$3'
∘ • • • ∘ ∘ ∘ ∘ ∘
3'$\underset{28}{C}$CAC$\underset{24}{A}$AUCAU$\underset{19}{}$5'   ψ**BP**

Figure 3: Structural model of the ψBP, the ψ-modified branch site duplex. A24, the branch site adenosine, is in an extrahelical position. The base triple between A24 and the A7-U22 Watson-Crick pair is shown to the upper left, and the 2'OH pointing into the major groove is shown in the lower left. Over-winding of the helix is shown in the view down the axis. Sequences of the strands are shown in the lower right.

The net result of the sharp kink is that the branch site adenine base (A24) is extruded from the helix. The Watson-Crick face of the base reaches upstream (with respect to the intron strand), and its major groove edge forms close contacts with the minor groove of the helix. Hydrogen bonds between A24 H62 amino proton and the N3 of A7, two base pairs upstream, parallel those of the G-A pseudo-pair in a GNRA tetraloop[23]. Although minor groove-minor groove interactions appear to be a common motif in the ribosome (the "A-minor motif")[24], the data do not support such an interaction here. Our structures indicate a nearly coplanar base triple between A24 and the A7-U22 Watson-Crick pair (Figure 3). ψ6 and A23 partially overlap in most structures instead of forming a canonical base pair, with no consensus as to which base stacks closer to G5. The G5-C25 base pair downstream of the branch site adenosine is of normal Watson-Crick geometry.

Additional support for the extrahelical position of the branch site adenine came from measurements of fluorescence of 2-aminopurine (2AP) in duplexes in which the fluorophore replaced adenine. 2AP fluorescence is quenched by the adjacent stacking of bases and not by hydrogen bonding[25] and, like A, 2AP can form two hydrogen bonds with an opposing U. 2AP fluorescence in uBP was similar to that of the probe in a single intron strand, yet several times greater than in the complementary duplex, implying partial stacking in the unmodified branch site duplex. Fluorescence of 2AP in the ψBP, however, was almost twice that of uBP and six times greater than in the complementary duplex, providing convincing evidence that the branch site adenosine adopts an extrahelical orientation in the ψ-modified duplex[16].

Adenosine residues appear unpaired more often than other nucleosides, possibly because adenine is the only base that cannot form three hydrogen bonds and it doesn't have a carbonyl oxygen, *i.e.* a strong electron-withdrawing group. Solution structures often depict extra adenosines stacked within the helix[22,26,27], presumably because of the energetic cost of solvating the relatively hydrophobic base in polar solvent. Exceptions involve stacking or hydrogen bonding in the minor groove, and include an extrahelical adenosine in the hairpin loop of the SL1 RNA of *C. elegans*[28] and crystallization of a model branch site duplex[29].

This local conformation is consistent with biochemical genetic data demonstrating accessibility of functional groups of the adenine base[30]. Moreover, the N1 position of the branch site adenosine, which has been proposed to function as a hydrogen bond acceptor in the first step of splicing[30], is prominently exposed in the minor groove in ψBP.

ψ-induced backbone deformation in the branch site helix

In addition to the local changes associated with the ψ-dependent structure, backbone deformation may contribute to formation of recognition elements in the splicing process. The observed over-winding of the helix (Figure 3) is the result of a lateral excursion of the intron strand backbone at the level of the extrahelical adenosine. The net result is an offset of the helical axis in the ψ-modified duplex between the stems flanking the branch site base, as compared with a single axis through the duplex of the unmodified counterpart (Figure 4).

The distortion of the backbone and base pairing in this structural motif has implications on the electrostatic profile of the major groove at the level of the branch site. Electrostatic calculations suggest a region of enhanced electronegativity in the major groove in the vicinity of the 2'OH that is not observed in the unmodified analogue (D. Xu, M.O. Fenley, N.L. Greenbaum, unpublished results).

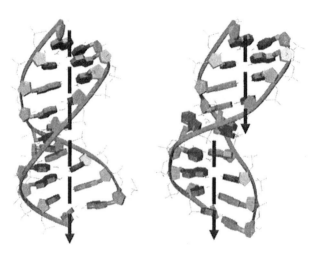

Figure 4: Schematic view of branch site helices in the unmodified (left) and ψ-modified (right) duplexes. RNA duplexes were rendered in hsc mode using DINO[31]. Presence of the modified base results in extrusion of the branch site adenosine into the minor groove, which, in turn, results in marked deformation of the helical structure seen in the unmodified helix. The axis of the unmodified helix is shown by a single arrow. The axis of the ψ-modified duplex, however, appears to be offset at the level of the branch site residue as a result of the overwinding of the intron strand. *(See page 12 of color insert.)*

Exposure of the 2'OH nucleophile

Significantly for biological function, the chain reversal also exposes the 2'OH of A24 to the major groove of the helix (Figure 3). Despite the apparent flexibility of the backbone in the region of the bulged base, the position of residues forming the base triple that anchors A24 was well fixed and the orientation of the 2'OH of A24 is essentially identical in all structures. The spatial context of the 2'OH provides a structural basis for recognition and access to the nucleophile by the RNA substrate strand in the first step of splicing.

Water-ψ interactions in the pre-mRNA branch site helix

The question arose how ψ, which differs from U primarily in the presence of an additional NH group, favors formation of such a markedly different structure with implication on function. The enhanced thermal stability noted in RNA duplexes containing ψ has been attributed to additional stabilization by a water-mediated hydrogen bond between the additional imino group and an oxygen atom of its own (5') phosphate[4], or by improved base stacking of ψ residues[8]. In all our structural models, the ψNH1, which resides in the major groove, is in line with its own phosphate oxygen (O1P) within an appropriate distance (3.6 ± 0.1 Å) to form a water-mediated hydrogen bond. Appearance of the ψNH1 resonance in a one-dimensional spectrum of exchangeable protons, and an NOE to the proton on the adjacent carbon (ψH6), implies that this proton is protected from exchange with solvent.

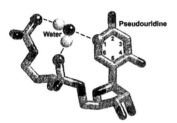

Figure 5. Schematic view of a water-mediated hydrogen bonding interaction between ψ and water. The water molecule is bridging an interaction between the ψNH1 group and phosphate oxygen atoms belonging to its own and the 5' nucleotide's phosphate group (adapted from ref. 32).

ROESY cross peaks between imino protons and water in the spectrum of the ψ-modified branch site duplex indicate exchange with water, but do not exclude contribution from an ROE interaction induced by cross-relaxation that is obscured by the exchange-induced ROE. To distinguish ROEs arising from pure chemical exchange from those having a cross-relaxation component, a CLEANEX-PM experiment[18] was utilized. This pulse sequence filters out cross-relaxation effects from tightly bound water molecules, which exchange much more slowly than the overall correlation time of the molecule, leaving only interactions between protons in the biomolecule and more loosely associated (more rapidly exchanging) waters. Thus, ROEs arising even partially from cross relaxation effects appears as a negative peak in a one-dimensional spectrum with an intensity proportional to specific characteristics of the interaction, including exchange rate, distance, and local motion. An NOE arising from pure chemical exchange will appear as a positive peak.

We first performed NOESY experiments to verify interaction between ψNH1 and water, followed by a ROESY experiment, from which we determined that the ψNH1-water interaction has a substantial element of chemical exchange.

We then applied a CLEANEX ROESY to evaluate the role of cross relaxation in the interaction. We observed a negative peak at 10.55 ppm, the resonance location corresponding to ψNH1, indicating cross-relaxation between ψNH1 and water protons. A similar observation was made for a complementary helix containing ψ, in which no conformational change is associated with presence of the modified base. For comparison, this experiment was also performed under identical conditions on the unmodified branch site duplex, which contains a U in place of ψ. No strong negative resonances were observed in the imino region for the unmodified duplex. The ψNH1 negative ROE builds up faster with increasing temperature. Broadening and disappearance of the negative ψNH1 resonance well below the predicted melting temperature suggests that the NOE between ψNH1 and water is dominated by chemical exchange at higher temperatures. Suppression of exchange relayed NOEs by the pulse sequence[18] and topology of the duplex[16] make it highly unlikely that other exchangeable protons contribute to the negative NOE we observe.

Taken together, these data from traditional ROESY and CLEANEX-PM spectroscopy unequivocally establish that the ψNH1 proton in the branch site RNA duplex is exchanging with the bulk solvent on a timescale faster than the correlation time of the molecule and is cross-relaxing with water, entirely consistent with a water-mediated hydrogen bond in the major groove.

Functional role of ψ in the branch site

The apparently strict conservation of this particular base modification argues strongly that it maintains an important functional role *in vivo*. Two lines of experimentation provide compelling evidence that this particular ψ modification (ψ35 in *S. cerevisiae*) favors formation of a structure that facilitates splicing: 1) formation of a catalytic product ('RNA X') by a protein-free complex of U2 snRNA, U6 snRNA, and a segment of the intron was greatly enhanced in the presence of this ψ[33]; and 2) genetic knockout of the pseudouridylating enzyme responsible for modifying this site in yeast, Pus 7, produced cells that were growth disadvantaged when grown in competition with wild-type cells (Y.-T. Yu, personal communication). As less direct evidence in agreement with this finding, replacement of this ψ with other bases (so long as Watson-Crick complementarity is maintained) results in cells in which splicing efficiency is decreased, but not annihilated[34]. We also speculate that the ψ-dependent structure assists in recognition of the branch site region by U2 snRNP splicing factors in assembly of the spliceosome.

Several lines of evidence suggest that A24 sometimes adopts an extrahelical position in uBP. The paucity of NOEs involving exchangeable protons in this region, and the broadness of A23 and A24 ribose proton resonances suggests conformational flexibility. As further evidence, we see a small, very broad peak in a one-dimensional spectrum at the chemical shift as A24H2 of ψBP. Moreover, the lower melting temperature of uBP than ψBP (by ~0.7 kcal/mole)[15], combined with fluorescence data suggesting at uBP A24 is not fully stacked, paints a dynamic picture of this region. Such conformational flexibility may explain why knockout of the pseudouridylase gene is not lethal and why certain base substitutions in the region are tolerated to some extent. It therefore appears that the role of the pseudouridine is to stabilize this extrahelical conformation, as one of many mechanisms utilized by the splicing machinery to facilitate correct branch site selection and to maintain branching efficiency.

A ψ has been identified at this specific position of U2 snRNA in every eukaryote investigated thus far, as well as in the human U12 snRNA of the *atac* spliceosome[35], and the structural features of ψBP are entirely consistent with all biochemical observations concerning branch site recognition and activity. We therefore consider it likely that the structural role of this ψ in the branch site motif seen in *S. cerevisiae* is a universal feature of eukaryotic spliceosomes and that, by helping to define a stable orientation for the branch site nucleophile, the branch site motif favored by ψ establishes a rationale for its phylogenetic preservation.

Analogous to our observations of nucleophile positioning in the spliceosomal pre-mRNA branch site, the exogenous guanosine nucleophile in Group I self-splicing introns is anchored by formation of a base triple[36], although it is positioned in the major groove. In contrast with these two systems is the conformation of the branch site nucleophile in a crystallographic structure of a construct of a Group II intron[37], which carries out the identical chemical mechanism as the spliceosome *in vitro* in the absence of proteins, and has no ψ. This model depicts the extrahelical branch site adenosine stacked between its 3'neighbor, a uridine, which is also extrahelical, and a base from a loop severed to aid crystallization. Further comparison of these systems will assist in determining other similarities and differences, which may help to identify whether the spliceosome is a ribozyme with properties similar to the self-splicing introns.

Acknowledgements

We acknowledge the NMR Facility, Department of Chemistry and Biochemistry, Florida State University, and the National High Magnetic Field Laboratory (Tallahassee, FL) for NMR facilities and support. We thank Darui Xu for creating Figure 4. M.N.L. was a recipient of a Molecular Biophysics fellowship funded by National Science Foundation Research Training Grant. This work was supported by NIH Grant GM54008 to N.L.G.

174

References

1. Moore, M., Query, C. & Sharp, P. In *The RNA World* (Gesteland, R. F. & Atkins, J. F., eds) Cold Spring Harbor Laboratory Press, Cold Spring Harbor, pp 303-357 (1993).
2. Madhani, H. D. & Guthrie, C. *Annu. Rev. Genet.* **28**, 1-26 (1994).
3. Lane, B. G. In *Modification and Editing of RNA* (Grosjean, H. & Benne, R., eds), ASM Press, Washington, DC, pp 1-20 (1998).
4. Davis, D. & Poulter, C. *Biochemistry* **30**, 4223-4231 (1991).
5. Hall, K. & McLaughlin, L. *Biochemistry* **30**, 1795-1801 (1991).
6. Durant, P. & Davis, D. *J. Mol. Biol.* **285**, 115-131 (1999).
7. Arnez, J. & Steitz, T. *Biochemistry* **33**, 7560-7567 (1994).
8. Yarian, C. S., Basti, M.M., Cain, R. J., Ansari, G., Guenther, R. H., Sochacka, E., Czerwinska, G., Malkiewicz, A., & Agris, P. F. *Nucleic Acids Res.* **27**, 3543-3549 (1999).
9. Auffinger, P. & Westhof, E. in *Modification and editing of RNA* (Grosjean, H., & Benne, R., eds), ASM Press, Washington, DC, pp 103-112 (1998).
10. Yu, Y.-T., Shu, M.-D. & Steitz, J. A. *EMBO J.* **17**, 5783-5795 (1998).
11. Reddy, R. & Busch, H. In *Structure and Function of Major and Minor Small Nuclear Ribonucleoprotein Particles*, (Birnstiel, M. L., ed), Springer-Verlag Press, Berlin, Germany, pp 1-37 (1988).
12. Patton, J., Jacobsen, M. & Pederson, T. *Proc. Natl. Acad. Sci. USA* **91**, 3324-3328 (1994).
13. Gu, J., Patton, J., Shimba, S, & Reddy, R. *RNA* **2**, 909-918 (1996).
14. Massenet, S. *et al. Mol. Cell. Biol.* **19**, 2142-2154 (1999).
15. Newby, M. I. & Greenbaum, N. L. *RNA* **7**, 833-845 (2001).
16. Newby, M. I. & Greenbaum, N. L. *Nature Structural Biology* **9**, 958-965 (2002).
17. Newby, M. I. & Greenbaum, N. L. *Proc. Natl. Acad. Sci.* **99**, 12697-12702 (2002).
18. Hwang, T.-L., Mori, S., Shaka, A. J. & van Zijl, P. C. M. *J. Am. Chem. Soc.* **119**, 6203-6204 (1997).
19. Rice, L. M. & Brünger, A. T. *Proteins* **19**, 277-290 (1994).
20. Stein, E. G., Rice, L. M. & Brünger, A. T. *J. Magn. Reson.* **124**, 154-164 (1997).
21. Brünger, A. T. *X-PLOR Version 3.851: A System for Crystallography and NMR*, Yale University Press, New Haven (1996).
22. Smith, J. S. & Nikonowicz, E. P. *Biochemistry* **37**, 13486-13498 (1998).
23. Jucker, F. M., Heus, H. A., Yip, P. F., Moors, E. H. M. & Pardi, A. A network of heterogeneous hydrogen bonds in GNRA tetraloops. *J. Mol. Biol.* **264**, 968-980 (1996).

24. Nissen, P., Ippolito, J.A., Ban, N., Moore, P.B. & Steitz, T.A. *Proc. Natl. Acad. Sci.,* 98, 4899-4903 (2001).
25. Rachofsky, E. L., Osman, R. & Ross, J. B. A. *Biochemistry* 40, 946-956 (2001).
26. Borer, P. N., *et al. Biochemistry* 34, 6488-6503 (1995).
27. Lynch, S. R. & Puglisi, J. D. *J. Mol. Biol.* 306, 1023-1035 (2001).
28. Greenbaum, N. L., Radhakrishnan, I., Patel, D. J. & Hirsh, D. *Structure* 4, 725-733 (1996).
29. Berglund, J. A., Rosbash, M. & Schultz, S. C. *RNA* 7, 682-691 (2001).
30. Query, C., Strobel, S. & Sharp, P. *EMBO J.* 15, 1392-1402 (1996).
31. Philippsen, A. DINO: Visualizing Structural Biology (http://www.dino3d.org)
32. Charette, M. & Gray, M. W. *IUBMB Life* 49, 341-351 (2000).
33. Valadkhan, S. & Manley, J.L. *RNA* 9, 892-904 (2003).
34. Wu, J. & Manley, J. L. *Genes Dev.* 3, 1553-1561 (1989).
35. Massenet, S. & Branlant, C. *RNA* 5, 1495-1503 (1999).
36. Kitamura, A. *et al. RNA* 8, 440-451 (2002).
37. Zhang, L. & Doudna, J. A. *Science* 295, 2084-2087 (2002).

Chapter 9

Structural Perturbations in Disease-Related Human tRNAs

Shana O. Kelley*, Lisa Wittenhagen, and Marc D. Roy

Boston College, Eugene F. Merkert Chemistry Center,
Chestnut Hill, MA 02467
*Corresponding author: shana.kelley@bc.edu

Genetic mutations affecting the sequences of human mitochondrial transfer RNAs (tRNAs) are implicated in a variety of degenerative diseases. The genes encoding hs mt tRNA^{Ile} and tRNA$^{Leu(UUR)}$ contain the highest number of pathogenic mitochondrial mutations. In our laboratory, pronounced changes in the structural and functional properties of disease-related mutants of both of these tRNAs have been detected. The observed losses in activity are often directly correlated with the disruption of structural elements. Many pathogenic mutations affect bases involved in the stabilization of secondary and tertiary structure, and the unique sequences of the hs mt tRNAs that produce weak structures appear to increase vulnerability to mutation-induced conformational defects. Therefore, the structural perturbations induced by disease-related tRNA mutations may constitute a molecular basis for pathogenicity.

Transfer RNAs bearing amino acids are essential components of the protein synthesis machinery (*1*). High-fidelity reactions catalyzed by the aminoacyl-tRNA synthestases (aaRSs) install amino acids onto the termini of tRNAs (*2*), and the tRNA anticodon/mRNA codon base-pairing interactions permit these residues to be placed within a polypeptide chain at genetically programmed positions. Many of the reactions and recognition events that involve tRNAs require specific structural or atomic features; thus the primary, secondary, and tertiary structures of tRNAs must all be intact for biological function.

An emerging class of diseases correlated with mutations in tRNAs encoded by the human mitochondrial (hs mt) genome constitute the first link between human pathologies and tRNA function (*3*). There are over 80 known pathogenic mutations within hs mt tRNA genes, and new mutations are continuously being identified (*4*). The types of diseases related to tRNA mutations are strikingly varied; disorders with symptoms affecting the neurological, cardiac, and endocrine systems have been reported in addition to those affecting hearing and sight.

Table 1. *Structural features and pathogenic mutations in hs mt tRNAs.*

Structural feature	nucleotides (nts) in conventional tRNAs	nts in hs mt tRNAs	pathogenic mt tRNA mutations
acceptor stem (a)	14	14	14
connector (b)	2	1-4	1
D stem (c)	8	0-8	10
D loop (d)	8-10	0-10	6
anticodon stem (e)	10	8-10	19
anticodon loop (f)	7	7-9	8
variable region (g)	4-24	3-5	5
TΨC stem (h)	10	10	6
TΨC loop (i)	7	3-9	10

ªFrom reference 5. ᵇFrom reference 7. ᶜFrom reference 4.

As shown in Table 1, the disease-related tRNA mutations appear in every domain of the cloverleaf structure typically used to describe the secondary structure of these molecules (*5*). Therefore, a variety of conformational defects likely result. The disruption of base-pairing interactions within stem regions or the loss of tertiary contacts that stabilize the three-dimensional fold of the tRNA would be expected to interfere with function. An *a priori* assessment of the effects of the pathogenic mutations on structure and function is difficult, however, because the sequences and structures of this class of tRNAs deviate significantly from the bacterial and cytoplasmic tRNAs that have been the focus of most studies in this field (*6, 7*). The limited number of investigations reported to date indicate that the destabilizing structural features of hs mt tRNAs contribute to the deactivating effects of pathogenic mutations (*8-15*). Therefore, understanding the structure-function relationships that govern the behavior of the tRNAs present within human mitochondria is useful in elucidating their involvement in specific disease states.

The Unique Structures of Mitochondrial tRNAs

A comparison of the sequences and structures of most tRNAs reveals an universal base-pairing pattern that forms a cloverleaf structure (*5*). In addition, highly conserved nucleotides can be identified that make interdomain hydrogen-bonding contacts stabilizing the L-shaped tertiary structure typical of tRNAs. However, the sequences of mt tRNAs, while still amenable to folding into a modified cloverleaf pattern, do not contain many of the conserved bases, and the size of the elements within the secondary structure is highly variable (*7*). Table 1 lists the number of nucleotides in each domain of the tRNA secondary structure. The comparison of the size of each structure for hs mt tRNAs and bacterial tRNAs reveals that the mt analogues generally feature significantly smaller domains. While bacterial tRNAs typically contain 74-95 nucleotides overall (*16*), the hs mt tRNAs have 62-78 nucleotides (*6*). In some cases, mt tRNAs are missing entire domains (*6*). It is interesting that these truncated tRNAs are functional in protein synthesis, as a uniform L-shaped structure is thought to be essential for efficient translation. The shortened tRNAs must form an analogous structure. The exact folding patterns for these sequences have not yet been elucidated, but computational and structural probing studies have identified sets of contacts that could produce stable tertiary structures (*17-20*). A recent review discusses mt tRNA structures in detail (*7*).

In addition to having sequences that are shorter and lack many of the nucleotides important for tertiary structure in other tRNAs, the base composition of hs mt tRNAs differs. In hs mt tRNAs, 33 to 75% of the Watson-Crick pairs are AU (*6*), while *E. coli* tRNAs, for example, contain between 11 and 38% AU pairs (*16*). The marked increase in the number of AU pairs, which are less stable than GC pairs because of one less hydrogen bond, is expected to increase the conformational lability of the hs mt tRNAs.

The few studies of the structural properties of hs mt tRNAs reported (*21-23*) indicate that the features mentioned above do give rise to lowered stability or more loosely structured conformations. One such study of hs mt tRNA^{Lys} (*23*) monitored the interdomain angle between the two helical regions of the folded L-structure using transient electric birefringence. Instead of the 90° angle found in more standard tRNAs, the mt tRNA displayed an angle of ~ 140°, indicating that the structure was significantly more open. Moreover, chemical and enzymatic cleavage of the same tRNA revealed that the absence of one modified base, 1-methyladenine, favored the formation of a misfolded conformer (*21-22*). These studies suggested that the hs mt tRNA^{Lys} structure was unstable, and that the absence of a single methyl group was sufficient to promote misfolding.

Impact of Disease-Related Mutations on the Structure And Function of Human Mitochondrial tRNAs

Clinical studies identifying mt tRNA mutations in patients exhibiting symptoms of mitochondrial disease have provided the first link between tRNA function and physiological function (*3*). In addition, the analysis of cell lines featuring the pathogenic mutations have confirmed that the substitutions cause detectable losses in respiration and protein synthesis (*21-24*). However, until recently, the effect of the mutations on the structures and functions of hs mt tRNAs remained unaddressed. In the last five years, a number of studies of pathogenic tRNA mutants have revealed that molecular-level defects could be detected and characterized (*8-15*).

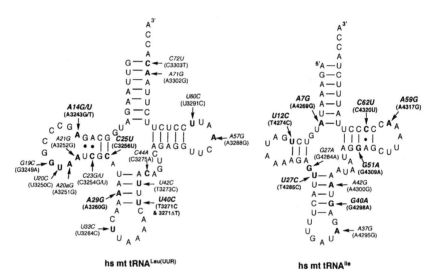

hs mt tRNA$^{Leu(UUR)}$

hs mt tRNA^{Ile}

Figure 1. Secondary structures of hs mt tRNA$^{Leu(UUR)}$ and tRNA^{Ile} illustrating positions of pathogenic mutations. The nucleotides are numbered according to the conventional system for tRNAs, and the corresponding genomic numbering is shown in parentheses. The mutations that are discussed in this chapter are bolded.

Hs mt tRNA^{Ile}

The gene encoding the tRNA carrying isoleucine in human mitochondria contains a high concentration of disease-related mutations (Figure 1) (*4*), with ten identified at present that correspond to different types of cardiomyopathies and opthalmoplegias. Several of these base substitutions have been monitored at the molecular level (*8, 9, 12, 15*), and the results of *in vitro* studies indicate that the unstable structure of this tRNA promotes functional deactivation in disease-related mutants. The hs mt tRNA^{Ile} contains all of the canonical structural domains, but has mismatched base pairs and high AU content that appears to render the structure weak.

Of the ten pathogenic mutations within hs mt tRNA^{Ile}, four produce CA mispairs within helical regions. The effect of three of these substitutions, U4274C (U12C), U4285C (U27C), and G4298A (G40A) on *in vitro* aminoacylation by hs mt IleRS was tested to determine whether function was disrupted by the mispairs (*8*). The aminoacylation efficiency of all three mutants was significantly reduced, with 25-1000-fold decreases in k$_{cat}$. The relationship of the lowered k$_{cat}$ values to structural defects was addressed with tRNA constructs containing compensatory mutations that changed the identity of the base pairs within the secondary structure but restored Watson-Crick pairing (Figure 2). The reintroduction of base pairing completely rescued activity, indicating that the loss of function for the pathogenic mutants was directly related to structural destabilization.

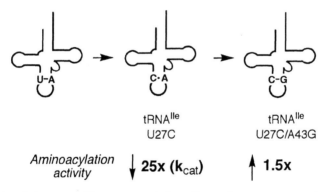

Figure 2. Schematic illustration of the effect of mutations producing CA mispairs in hs mt tRNA^{Ile}. For example, the CA pair brought about by the opthalmoplegia-related U27C (U4285C) mutation decreases aminoacylation efficiency by reducing catalytic turnover. The functional defect results from a loss of structure in the anticodon stem, as the incorporation of a second mutation restoring base pairing completely rescues the activity. The arrows and changes in K_m and k_{cat} shown are relative to the WT tRNA. Adapted from ref. 8.

A structurally weak TΨC stem within hs mt tRNA^{Ile} also appears to amplify the deleterious effects of pathogenic mutations. *In vitro* studies of mutations in this region suggested that the TΨC stem can undergo deactivating rearrangements (*9*). The A4317G (A59G) mutation affecting an unpaired base which replaces a CA pair with an UA pair within the appended stem. The A4317G mutation decreased the reactivity, while the stabilizing C4320U substitution increased the charging efficiency. When the A4317G mutation was introduced into a tRNA also containing the C4320U substitution (Figure 3), reactivity was not affected, indicating that the latter stabilizing mutation prevented a rearrangement caused by A4317G. These results and the analysis of other mutants with altered base pairing arrangements in the TΨC stem indicated that the CA mispair imbedded within this domain promotes misalignment of the helix. The misfolded tRNA appears to be a poor substrate for IleRS.

The weak TΨC stem augments the impact of other pathogenic mutations. Constructs containing the stabilizing C4320U mutation and destabilizing U4274C, U4285C, or G4298A mutations were tested, and the effects of the pathogenic mutations were less severe than in the WT tRNA[Ile] (9). This observation indicates that interdomain communication within this tRNA promotes the deleterious effects of pathogenic mutations by propagating thermodynamic weaknesses throughout the structure. It is noteworthy that most of the studies conducted to date have utilized unmodified tRNA transcripts to monitor the effects of pathogenic mutations. It is possible that the lack of modified bases may produce more flexible structures that are found *in vivo*. Therefore, it will be important to revisit similar studies using authentic tRNA samples isolated from biological sources.

A recent study by Levinger and coworkers indicated that 3'-processing is also affected by a subset of the hs mt tRNA[Ile] mutations that disrupt structure (*15*). The A4269G, G4309A, and A4317G mutations decreased the efficiency of processing ~10-fold, illustrating that the effects of disease-related mutations

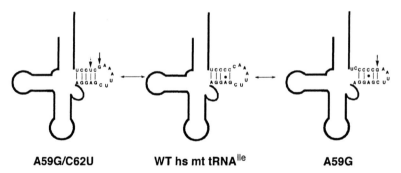

A59G/C62U   **WT hs mt tRNA[Ile]**   **A59G**

Figure 3. The A59G (A4317G) mutation (indicated by solid arrow) shifts the alignment of the TΨC stem and reduces aminoacylation efficiency (right). In the presence of a C62U (C4320U) mutation (indicated by dotted arrow) that stabilizes the stem, the effects of the A59G substitution are minimal (left). Adapted from ref. 9.

may be reflected in many different aspects of tRNA function.

The human mitochondrial tRNA^{Ile} appears vulnerable to functional deactivation by disease-related mutations because of its intrinsic structural fragility. The high AU base content (~70%) and a CA mispair within the TΨC stem limits the ability of this molecule to retain the correct fold when the primary sequence is altered by the introduction of pathogenic mutations. Therefore, the involvement of this tRNA in human disease may be directly related to its susceptibility to structural perturbations.

Hs mt tRNA$^{Leu(UUR)}$

Eighteen mutations within the hs mt tRNA$^{Leu(UUR)}$ gene have been correlated with various diseases (Figure 1) (4). Additionally, a subset of the substitutions found in this gene are the most prevalent tRNA mutations detected in human subjects. Because tRNA$^{Leu(UUR)}$ contains the highest concentration of disease-related tRNA mutations, most of the studies focused on identifying cellular phenotypes correlated with the pathogenic genotypes have focused on this system (24-29). Only recently, however, have the impact of mutations on molecular structure of hs mt tRNA$^{Leu(UUR)}$ been addressed (10, 11, 14).

An A3243G (A14G) substitution within hs mt tRNA$^{Leu(UUR)}$ is the most common mutation within the mt genome detected in clinical studies (30). Found in significant numbers of diabetes cases in certain geographic areas, A3243G is also associated with MELAS (mitochondrial myopathy, encephalopathy, lactic acidosis, and stroke-like episodes), cardiomyopathy, and opthalmoplegia. A number of different cell lines harboring this mutation have been generated, but many disparate conclusions have been drawn from studies at the cellular level concerning the exact defect associated with the A3243G substitution (24-29). Differing translational efficiencies have been observed in independent studies. In addition, while decreased overall levels of the tRNA$^{Leu(UUR)}$ have been reported by several groups, the extent to which the mutated tRNA is aminoacylated in cells has varied in different systems (31).

A recent study of the A3243G mutant of hs mt tRNA$^{Leu(UUR)}$ affecting A14 within the D-loop revealed a dramatic structural change promoted by the presence of the mutation (10). Using native gel electrophoresis, a high-affinity (K_d ~ 150 nM) dimeric complex was detected for the mutated tRNA (Figure 4). A stretch of self-complementary nucleotides in the D-loop was identified that nucleates the complex assembly. The A3243G mutant exhibits diminished reactivity with hs mt LeuRS, and formation of the dimeric complex contributes to the loss of aminoacylation *in vitro*. In addition, the A3243G mutation disrupts a highly conserved contact between A14 and U8, thus the destabilization of the tertiary fold likely also interferes with efficient aminoacylation.

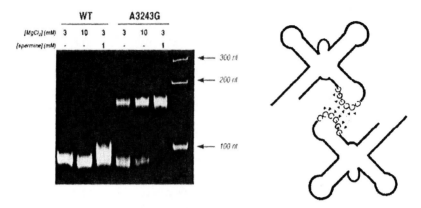

Figure 4. Dimerization of the A3243G tRNA^{Leu(UUR)} mutant monitored by native gel electrophoresis. The presence of the pathogenic mutation induces the formation of a dimeric complex, particularly in the presence of MgCl_2 or spermine. Adapted from ref. 10.

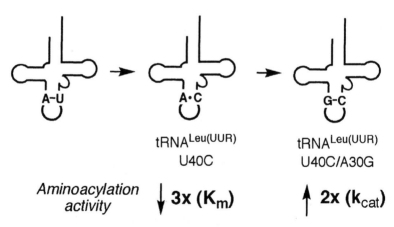

Figure 5. Schematic illustration of the effect of mutations producing CA mispairs in hs mt tRNA^{Leu(UUR)}. Interestingly, the same trends are observed as for the anticodon mutant shown in Figure 2, with the destabilizing mispair diminishing activity (albeit here by increasing K_m) and the introduction of a stabilizing mutation mitigating the effect. The arrows and changes in K_m and k_{cat} are relative to the WT tRNA. Adapted from ref. 11.

An U3271C (U40C) mutation within hs mt tRNA$^{Leu(UUR)}$ associated with MELAS and diabetes has also been investigated (*11, 14*). Analogous to the tRNA^{Ile} mutants discussed above, this tRNA$^{Leu(UUR)}$ variant has a CA mismatch in the anticodon stem. Because the hs mt tRNA$^{Leu(UUR)}$ anticodon stem contains only AU pairs, it is likely even weaker than in hs mt tRNA^{Ile}.

Studies in our laboratory (*11*) indicated that the U3271C hs mt tRNA$^{Leu(UUR)}$ mutation decreased aminoacylation efficiency, but the effect was more subtle compared to the tRNA^{Ile} mutants discussed above (here, a 3-fold decrease in efficiency for the tRNA$^{Leu(UUR)}$ mutant was observed compared to a 25-1000-fold decrease for the tRNA^{Ile} mutants). In addition, while decreased aminoacylation efficiencies for tRNA^{Ile} mutants with mismatches in the anticodon stem resulted from a reduction of k_{cat}, the U3271C tRNA$^{Leu(UUR)}$ mutant was more poorly aminoacylated because of an increased K_m (Figure 5). Chemical probing experiments confirmed that a pronounced disruption was induced by the U-to-C substitution, and that the structure of the stem was completely restored by the introduction of a compensatory mutation. This double mutant also displayed restored reactivity, with increased catalytic efficiency as compared to the WT tRNA because of a k_{cat} increase. Thus, while the mutations in tRNA$^{Leu(UUR)}$ and tRNA^{Ile} may impact the integrities of these structures similarly, the degree to which function is affected differs, likely because the enzymes that react with these tRNAs recognize the substrates using unique sets of nucleotides.

The A3243G and U3271C mutations within hs mt tRNA$^{Leu(UUR)}$ were also examined by King and coworkers (*14*) using tRNA substrates isolated from human cells. While an even larger decrease in aminoacylation efficiency was observed for the native A3243G tRNA$^{Leu(UUR)}$ as compared to an *in vitro* transcript of the same tRNA, the effect of the U3271C mutation on activity was less pronounced. These results illustrate that modified bases do have an impact on the activity of hs mt tRNAs, and demonstrate the value of isolating authentic tRNA samples.

Processing defects caused by disease-related mutations in hs mt tRNA$^{Leu(UUR)}$ have also been detected (*13*). In particular, the C3256U, A3260G, and U3271C mutations, all producing mispairs in helical regions, decreased the processing efficiencies by 5-20 fold. The A3243G mutation also attenuated activity. Structural defects, whether affecting secondary or tertiary structure, may therefore limit tRNA levels by prohibiting the cleavage reactions that are essential for generation of the native sequence.

Conclusions

Studies focused on the molecular-level properties of disease-related mutants of hs mt tRNAs have repeatedly revealed that structural integrity is disrupted. A variety of specific functional consequences have been correlated to these conformational defects, but it is reasonable to expect that the mutated tRNAs with altered structures will not function properly in any aspect of protein synthesis. Therefore, changes in secondary, tertiary, and even quaternary structure induced by single-base changes in primary sequence may contribute to the manifestation of diseases caused by mt tRNA mutations. The weak structures of these tRNAs, with domains destabilized by high AU content and mismatched base pairs, appear highly vulnerable to structural perturbations. By formulating a detailed understanding of the structure-function relationships that govern the activities of these novel tRNAs, we hope to further understand how human diseases may be connected to the molecular properties of these important cellular components.

Literature Cited

1. *Transfer RNA: Structure, Properties, and Recognition*; Schimmel, P.; Söll, D.; Abelson, J., Eds.; Cold Springs Harbor: USA, 1979.
2. Schimmel, P. in *Transfer RNA: Structure, Properties, and Recognition;* Schimmel, P.; Soll, D.; Abelson, J., Eds.; Cold Springs Harbor: USA, 1979; pp. 297-310.
3. DiMauro, S.; Schon, E. A. *Am. J. Med. Genet.* **2001,** *106*, 18-26.
4. MITOMAP: A Human Mitochondrial Genome Database. Center for Molecular Medicine, Emory University, Atlanta, GA, USA. (2000), http://www.mitomap.org/.
5. Dirheimer, G.; Keith, G.; Dumas, P.; Westhof, E. in *tRNA Structure, Biosynthesis, and Function;* Soll, D.; RajBhandary, U., Eds.; ASM Press: Washington, D.C., 1995; pp. 93-126.
6. Sprinzl, M.; Horn, C.; Brown, M.; Ioudovitch, A.; Steinberg, S. *Nucl. Acids Res.* **1998,** *26*, 148-153.
7. Helm, M.; Brule, H.; Friede, D.; Giegé, R.; Putz, D.; Florentz, C. *RNA* **2000,** *6*, 1356-1379.
8. Kelley, S. O.; Steinberg, S. V.; Schimmel, P. *Nat. Struct. Biol.* **2000,** *7*, 862-865.
9. Kelley, S. O.; Steinberg, S. V.; Schimmel, P. *J. Biol. Chem.* **2001,** *276*, 10607-10611.
10. Wittenhagen, L. M.; Kelley, S. O. *Nat. Struct. Biol.* **2002,** *9*, 586-590.
11. Wittenhagen, L. M.; Roy, M. D.; Kelley, S. O. *Nucl. Acids Res.* **2003,** *31*, 596-601.

12. Degoul, F.; Brule, H.; Cepanec, C.; Helm, M.; Marsac, C.; Leroux, J.; Giegé, R.; Florentz, C. *Hum. Mol. Genet.* **1998**, *7*, 347-354.

13. Rossmanith, W.; Karwan, R. M. *FEBS Lett.* **1998**, *433*, 269-274.

14. Park, H.; Davidson, E.; King, M. P. *Biochemistry* **2003**, *42*, 958-964.

15. Levinger, L.; Giegé, R.; Florentz, C. *Nucl. Acids Res.* **2003**, *31*, 1904-1912..

16. Gauss, D.; Gruter, F.; Sprinzl, M. in *Transfer RNA: Structure, Properties, and Function;* Schimmel, P.; Söll, D.; Abelson, J., Eds.; Cold Springs Harbor: USA, 1979; pp. 520-537.

17. De Bruijn, MH; Klug, A. *EMBO J.* **1983**, *2*, 1309-21.

18. Steinberg, S.; Leclerc, F.; Cedergren, R. *J. Mol. Biol.* **1997**, *266*, 269-282.

19. Yokogawa, T; Watanabe, Y.; Ueda, T.; Hirao, I.; Miura, K.; Watanabe, K. *Nucl. Acids Res.* **1991**, *19*, 6101-6105.

20. de Bruijn, M. H.; Schreier, P. H.; Eperon, I. C,; Barrell, B. G.; Chen, E. Y.; Armstrong, P. W.; Wong, J. F.; Roe, B. A. *Nucl. Acids Res.* **1980**, *8*, 5213-5222.

21. Helm, M.; Brule, H.; Degoul, F.; Cepanec, C.; Leroux, J. P.; Giegé, R.; Florentz, C. *Nucl. Acids Res.* **1998**, *26*, 1636-1643.

22. Helm, M.; Giegé, R.; Florentz, C. *Biochemistry* **1999**, *38*, 13338-13346.

23. Leehey, M. A.; Squassoni, C. A.; Friederich, M. W.; Mills, J. B.; Hagerman, P. J. *Biochemistry* **1995**, *34*, 16235-16239.

24. King, M. P.; Koga, Y.; Davidson, M.; Schon, E. A. *Mol. Cell. Biol.* **1992**, *12*, 480-490.

25. Chomyn, A.; Martinuzzi, A.; Yoneda, M.; Daga, A.; Hurko, O.; Johns, D.; Lai, S. T.; Nonaka, I.; Angelini, C.; Attardi, G. *Proc. Natl. Acad. Sci. USA* **1992**, *89*, 4221-4225.

26. Schon, E. A.; Koga, Y.; Davidson, M.; Moraes, C. T.; King, M. P. *Biochim. Biophys. Acta* **1992**, *1101*, 206-209.

27. El Meziane, A.; Lehtinen, S. K.; Hance, N.; Nijtmans, L. G.; Dunbar, D.; Holt, I. J.; Jacobs, H. T. *Nat. Genet.* **1998**, *18*, 350-353.

28. Janssen, G. M.; Maassen, J. A.; van den Ouweland, J. M. *J. Biol. Chem.* **1999**, *274*, 29744-29748.

29. Flierl, A.; Reichmann, H.; Seibel, P. *J. Biol. Chem.* **1997**, *272*, 27189-27196.

30. Maassen, J. A. *Am. J. Med. Genet.* **2002**, *115*, 66-70.

31. Jacobs, H. T.; Holt, I. J. *Hum. Mol. Genet.* **2000**, *9*, 463-465.

Chapter 10

Folding of Bent Oligodeoxynucleotide Duplexes: Energetics, Ion Binding, and Hydration

Luis A. Marky[1,2], Karen Alessi[1], Ronald Shikiya[1], Jian-Sen Li[2], and Barry Gold[1,2]

[1]Department of Pharmaceutical Sciences and [2]Eppley Institute for Research in Cancer, University of Nebraska Medical Center, Omaha, NE 68198–6025

In this review, we focus on the overall thermodynamics of the folding (or unfolding) of DNA oligomer structures that are distorted by the inclusion of small molecules. Specifically, we concentrate on the analysis of the energetics involved in the physical and chemical placement of ligands into DNA, and their correlations with volume changes that shed light on specific hydration contributions to DNA bending. Four different systems are discussed: i) helix-coil transition of oligomer duplexes containing dA•dT tracts and their interaction with the minor groove ligand netropsin; ii) the covalent placement of a benzopyrene derivative into duplex DNA; iii) the covalent placement of cisplatin into a decamer duplex; and iv) the inclusion of aminopropyl cationic chains into the Dickerson-Drew dodecamer. The presence of a bend in these systems has been confirmed indirectly by gel mobility studies or directly by NMR and X-ray crystallographic investigations. The presence of a bend in a DNA duplex is accompanied by a loss of favorable stacking interactions yielding a greater exposure of polar and non-polar groups to the solvent. This greater exposure of uncharged groups is further stabilized by the concomitant immobilization of structural water. The overall effect correlates qualitatively with calculations of the solvent accessible surface areas.

INTRODUCTION

DNA is normally considered to be a stiff rod-like molecule with a persistent length of over 100 base pairs (*1*). It has been suggested that DNA remains rigid and linear because any significant deviation from this basic structure causes unfavorable disruption of base stacking due to hydrophobic effects (*2-4*) and electrostatic repulsion between the anionic charges of the intra-strand phosphate groups (*1,5*) (Figure 1). Proteins and small molecules, which are physically or chemically bound to DNA, would overcome the barrier(s) to non-linearity by changing the local environment of DNA. Deviations from nonlinear DNA structures can include static bends or kinks or dynamic isotropic and anisotropic flexibility. We define all of these as distorted DNA structures because most experimental methods that are used to measure "bending" actually measure the average conformation of populations of molecules and cannot distinguish between static and dynamic states.

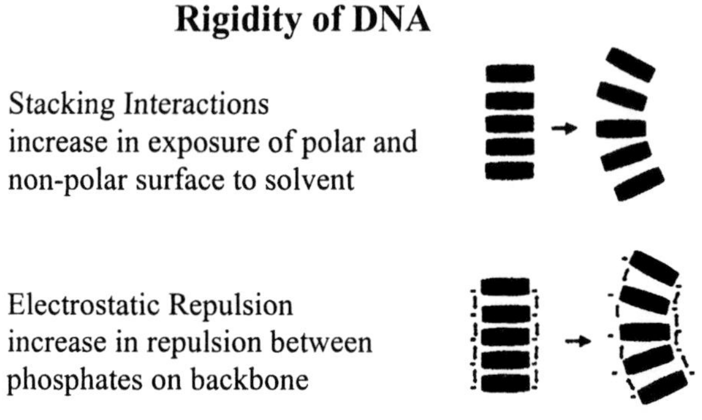

Rigidity of DNA

Stacking Interactions
increase in exposure of polar and
non-polar surface to solvent

Electrostatic Repulsion
increase in repulsion between
phosphates on backbone

Figure 1. Examples of physical factors that distort the rigid DNA helix.

There are numerous structures of DNA-protein complexes in which the DNA component is nonlinear (*6*). The complexes can be divided into two categories: those in which the protein associates with DNA predominantly *via* major groove interactions and those in which the protein contacts DNA from the minor groove. In the former, the DNA adopts a shape that wraps around the protein, while in the latter the DNA is bent away from the protein. Using an internal perspective, in both scenarios the DNA has the same trajectory in space. The exact relationship between the physical shape(s) of DNA and biochemical functionality is not often clear. It has been suggested that deformation of DNA by a protein may enhance binding by a required accessory protein or that DNA bending may place remote DNA sites in close proximity as required by looping models of transcriptional regulation. In fact, these two outcomes are not mutually exclusive. An obvious case where the biologic importance of protein-induced DNA bending is clear is in the formation of nucleosomes, which is required for DNA packaging inside the cell (*7*).

Water plays an important role in the conformational stability of nucleic acids (*8, 9*). The overall hydration of a nucleic acid depends on the base sequence, composition and conformation (*10-12*). However, to investigate the hydration of a nucleic acid is a challenging problem because of the low occupancy exchange and presence of at least two types of hydrating water, electrostricted (around charged groups) and structural or hydrophobic (around uncharged groups i.e., polar and non-polar groups) (*13-15*). The volume effects accompanying the formation of a nucleic acid, from mixing its complementary strands, are interpreted to reflect changes in the electrostriction and/or hydrophobicity of water dipoles, which are immobilized by each of the participating species. The volume change for an association reaction is a reflection of the differences in the molar volume of immobilized water relative to bulk water. For instance, the single strands have the bases more exposed to the solvent and immobilize more structural water. The base pairing and base pair stacks of the duplex reduce the exposure of the bases to solvent, yielding a duplex with a higher charge density that interacts with counterions more effectively and immobilizes more electrostricted water. The contribution of the release (or uptake) of counterions to the overall volume change in these association reactions is considered negligible because they keep their hydration shells intact i.e., their hydrated molar volume remains constant. In this light, the formation of a nucleic acid duplex is interpreted to reflect changes in electrostriction of water dipoles and changes of structurally "bound" water (*13-15*), effects that compensate for each other (*16-18*). However, our laboratory has obtained a physical signature for the identification of the primary type of water that is involved in the inclusion of perturbations into a nucleic acid duplex. This is done by making parallel measurements of standard thermodynamic profiles and volume changes, and establishing suitable thermodynamic cycles using proper controls or reference reactions. The comparison of the signs of the resulting differential free energy, or differential enthalpy-differential entropy compensation, with the differential volume change allows the distinction of the type of water that is involved in these thermodynamic cycles. Similar signs of the $\Delta\Delta G°$ (or $\Delta\Delta H$-$\Delta(T\Delta S)$ compensation) with the $\Delta\Delta V$ indicate that the differential hydration reflects electrostriction effects; opposite signs of these terms indicate the participation of structural water (*17,19-20*).

In this review, we focus on the overall thermodynamics of the folding (or unfolding) of DNA oligomer structures that are distorted by the inclusion of small molecules. Specifically, we do an analysis of the molecular forces associated with covalent and/or non-covalent modifications of DNA. The resulting volume changes shed light on specific hydration contributions to DNA bending. Four different systems are discussed: i) helix-coil transition of oligomer duplexes containing runs of dA•dT base pairs and their interaction with the minor groove ligand netropsin; ii) the covalent placement of a benzopyrene derivative into duplex DNA; iii) the covalent placement of cisplatin into a decamer duplex; and iv) the inclusion of aminopropyl cationic chains into the Dickerson-Drew dodecamer.

Binding of Netropsin to Model Bent Sequences.

Electrophoretic studies on oligonucleotide duplexes with the sequences $[d(GA_4T_4C)]_2$ and $[d(GT_4A_4C)]_2$ (*21*) (Figure 2), as well as the ligated sequences $(GA_4T_4C)_n$ and $(GT_4A_4C)_n$ with n > 5 (*22*) have shown that the $(GA_4T_4C)_n$ sequences have retarded gel mobility while the $(GT_4A_4C)_n$ sequences have normal mobility. NMR studies of the $[d(GA_4T_4C)]_2$ and $[d(GT_4A_4C)]_2$ duplexes also concluded that the former is bent, whereas the latter is not (*23*).

```
G-A-A-A-A-T-T-T-T-C
•  •  •  •  •  •  •  •  •  •
C-T-T-T-T-A-A-A-A-G

G-T-T-T-T-A-A-A-A-C
•  •  •  •  •  •  •  •  •  •
C-A-A-A-A-T-T-T-T-G
```

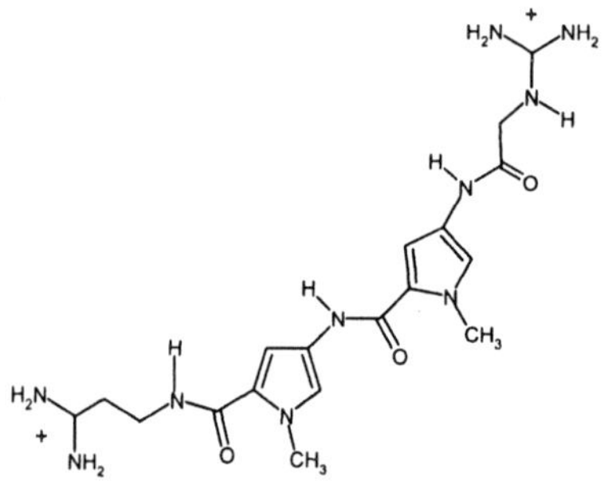

Figure 2. Sequences of Oligonucleotides and structure of netropsin.

In addition, spectroscopic and calorimetric investigations on these decamer duplexes indicated an unusual premelting phenomenon and anomalous heat capacity differences for only the [d(GA_4T_4C)]$_2$ duplex (*24*). Our UV and differential scanning calorimetry (DSC) melting studies (Figure 3) yielded the thermodynamic profiles shown in Table 1 (*25*). In 0.1 M NaCl, the [d(GA_4T_4C)]$_2$ duplex is thermally more stable by 3 °C and shows a higher exothermic enthalpy of -15.4 kcal/mol (Table 1). These two terms combined yield a favorable $\Delta\Delta G^\circ$ of -1.6 kcal/mol; therefore, the bent [d(GA_4T_4C)]$_2$ duplex is more stable than the [d(GT_4A_4C)]$_2$ duplex. The formation of the [d(GA_4T_4C)]$_2$ duplex also shows a higher uptake of counterions, Δn_{Na}^+, by 0.26 mol Na$^+$/mol duplex.

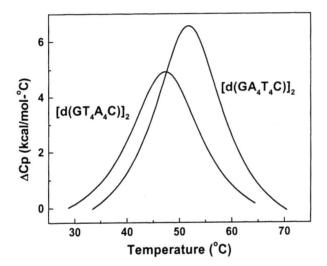

Figure 3. DSC melting curves of decamer duplexes in 10 mM Sodium phosphate buffer at pH 7 and 1 M NaCl.

Table 1. Thermodynamic Parameters of Duplex Formation at 20 °C.[a]

Duplex	T_M (°C)	ΔH_{DSC} (kcal/mol)	ΔG° (kcal/mol)	$T\Delta S$ (kcal/mol)	Δn_{Na}^+ (mol Na$^+$/mol)
[d(GA_4T_4C)]$_2$	40	-84.4	-5.4	-79.0	-3.01
[d(GT_4A_4C)]$_2$	37	-69.0	-3.8	-65.2	-2.75

[a]All parameters are calculated per mole of duplex and are measured in 10 mM sodium phosphate buffer at pH 7.0 and 0.1 M NaCl. The experimental uncertainties are as follow: T_M (±0.5 °C), ΔH_{DSC} (±3%), ΔG° (±4%), $T\Delta S$ (±5%) and Δn_{Na}^+ (5%).

If we assumed that at high temperature the conformation of the single strands is nearly identical; then, the differential Δn_{Na^+} term indicates that the bent duplex has a higher degree of helicity yielding a higher charge density parameter, which is consistent with its higher value of the folding enthalpy. In this assumption, the d(GA$_4$T$_4$C) and d(GT$_4$A$_4$C) strands are considered true random coils because of their similar base composition and at high temperatures the base stacking contributions are negligible. A preliminary conclusion is that the higher stability of the bent duplex is due to more favorable base-pair stacking contributions. However, closer inspection of the type of base-pair stacks in each duplex indicates that [d(GA$_4$T$_4$C)]$_2$ has 2 GA/CT, 6 AA/TT and 1 AT/TA base pair stacks while [d(GT$_4$A$_4$C)]$_2$ has 2 GT/CA, 6 AA/TT and 1 TA/AT base pair stacks. The enthalpy associated with this difference in base pair stacks, calculated from nearest-neighbor parameters, is equal to 0.8-2.2 kcal (26,27). We speculate that the higher value of the unfolding enthalpy of the bent duplex, (by 14.4 kcal) may be due to an additional enthalpy contribution from the higher immobilization of water molecules by the bent duplex, other contributions such as the ionic contributions contribute negligibly to the heat (18,28).

Electrophoretic studies performed on kinetoplast DNA with a variety of minor groove and intercalating ligands have shown that the interaction of ligands tends to straighten bent structures (29). Therefore, we studied the interaction of netropsin with each duplex to obtain complete thermodynamic binding profiles, the [d(GT$_4$A$_4$C)]$_2$ is used as a control duplex. We used UV melting techniques to obtain the associated binding affinities, K_b, (and binding free energies, $\Delta G°_b$) from the increase in thermal stability of the 1:1 and 2:1 complexes relative to the free duplexes. In addition, we used isothermal titration calorimetry (ITC) to obtain the interacting heats, ΔH_b; and volumetric techniques to measure the associated volume changes, ΔV_b (19,25,30) ITC titrations of each oligonuneotide with netropsin are shown in Figure 4 and the resulting and complete thermodynamic profiles are shown in Table 2. Each oligomer duplex binds two netropsin molecules with similar K_b's of ~3.6 x 10^7. Binding of each netropsin molecule to the [d(GT$_4$A$_4$C)]$_2$ is primarily enthalpy driven while the binding of the first netropsin molecule to [d(GA$_4$T$_4$C)]$_2$ is entropy driven and the binding of the second netropsin is both enthalpy and entropy driven. The latter results may be explained in terms of an endothermic hydration contribution which lowers the overall exothermicity of the intrinsic binding enthalpy, confirming our preliminary conclusion from the unfolding results.

The comparison of the resulting volume effects yielded the type of water involved in bent sequences. We obtained volume changes by mixing netropsin and oligomer duplex in two different ways: to form 1:1 ligand-oligomer complexes followed by the addition of a second equivalent of ligand, and to

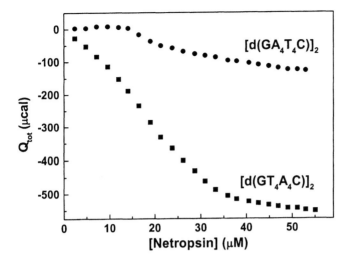

Figure 4. ITC binding curves of decamers duplexes in 10 mM sodium phosphate buffer at pH 7.0 and 0.1 M NaCl.

Table 2. Thermodynamic Profiles of Netropsin Binding to DNA at 20 °C.[a]

Duplex	Site	K_b (M^{-1})	$\Delta G°_b$ (kcal/mol)	ΔH_b (kcal/mol)	$T\Delta S_b$ (kcal/mol)	ΔV_b (ml/mol)
[d(GA₄T₄C)]₂	1st	2.2 x 10⁷	-9.9	1.2	11.1	-100
	2nd	2.2 x 10⁷	-9.9	-4.0	5.9	37
[d(GT₄A₄C)]₂	1st	4.9 x 10⁷	-10.3	-11.1	-0.8	-83
	2nd	4.9 x 10⁷	-10.3	-9.7	0.6	15

[a]All parameters are calculated per mole of duplex and are measured in 10 mM sodium phosphate buffer at pH 7.0 and 0.1 M NaCl. The experimental uncertainties are as follow: K_b (±35%), $\Delta G°_b$ (±7%), ΔH_b (±3%), $T\Delta S_b$ (±5%) and ΔV (±7 %).

form directly the 2:1 complexes; both methods yielded similar results shown in the last column of Table 2. Binding of netropsin to the first site of each oligomer duplex generated a volume contraction whereas a volume expansion is observed in the binding to the second site. The magnitude of these volume effects was higher with the [d(GA₄T₄C)]₂ oligomer than with its [d(GT₄A₄C)]₂ isomer. Thus, at this salt concentration, we infer that a net hydration accompanies the total binding to these oligomer duplexes as reflected by the apparent compression of water molecules. The $[d(GA_4T_4C)]_2$ is slightly more hydrated by 5 ml/mol of duplex and indicates that this net hydration depends on the DNA sequence and the type of site of the oligomer duplex. To obtain a better understanding of the observed differences, we have dissected the overall volume profiles into the individual contributions for netropsin binding to the first and second site, respectively, and have taken the differences for each site, using the formation of the netropsin-$[d(GT_4A_4C)]_2$ complex as the reference state. The resulting $\Delta\Delta V$ parameters are -17 ml/mol for the first site and 22 ml/mol for the second site. A similar procedure for the dissection of the other thermodynamic parameters yielded a marginal $\Delta\Delta G°$ of 0.4 kcal/mol for each binding site of netropsin; $\Delta\Delta H_b$ of 12.3 kcal/mol and 5.7 kcal/mol and $\Delta(T\Delta S_b)$ of 11.9 kcal/mol and 6.5 kcal/mol for the first and second binding site, respectively. These results indicate that in the sequential binding of netropsin, there is an initial increase in hydration due to a hydrophobic or structural effect, opposite signs of $\Delta\Delta V$ and $\Delta\Delta G°$ or $\Delta\Delta H_b$-$\Delta(T\Delta S_b)$, followed by a dehydration event due to electrostriction effects, similar signs of $\Delta\Delta V$ and $\Delta\Delta G°$. Analysis of the electrophoretic mobility (data not shown) of the free and netropsin-bound oligomer duplexes shows that the mobility of the free $[d(GA_4T_4C)]_2$ is lower than that of the free $[d(GT_4A_4C)]_2$; the 1:1 netropsin- oligomer complexes ran with similar mobilities whereas for the 2:1 complexes this trend is reversed (*25*). These results strongly indicate that binding of the first netropsin straightens the curved $[d(GA_4T_4C)]_2$ duplex, yielding an important correlation with our thermodynamic studies in that the straightening of this oligomer invokes the participation of structural water or, alternatively, the loss of electrostricted water.

Formation of Oligomer Duplexes with Covalently Attached Ligands.

In this section, we present results for two sets of oligonucleotides: the inclusion of two enantiomers of benzo[a]pyrene diol epoxide (BPDE), (+)-BPDE and (-)-BPDE, at the central guanine (boldface) of the duplex, d(CCATCGCTACC) /d(GGTAGCGATGG), forming a DNA adduct; and the covalent attachment of cisplatin, cis-$[Pt(NH_3)_2Cl_2]$, to the central **GG** step of d(CTCT**GG**TCTC)/d(GAGACCAGAG) and their control unmodified duplexes.

In order to obtain complete thermodynamic profiles for the formation of these DNA adducts at 20 °C, and to correlate the resulting energetics with the molecular interactions observed in their solution structures, we first used density and ITC techniques to measure the ΔV and ΔH_{ITC} of duplex formation, from the mixing of complementary strands, respectively. The additional $\Delta G°_{DSC}$, ΔH_{DSC} and $T\Delta S°_{DSC}$ parameters are determined from the standard thermodynamic profiles of the helix—coil transition of each duplex that are measured in DSC experiments. The $\Delta G°_{DSC}$ term is temperature extrapolated from the T_M to 20 °C and corrected for the contribution of disrupting base-base stacking interactions in the single strands at 20 °C to yield $\Delta G°_{(20)}$ i.e., $\Delta G°_{(20)} = \Delta G°_{DSC}$ $(\Delta H_{ITC}/\Delta H_{DSC})$. The $T\Delta S°_{(20)}$ is then calculated from the Gibbs equation: $\Delta G°_{(20)} = \Delta H_{ITC} - T\Delta S°_{(20)}$. We also used UV melting techniques (in conjunction with DSC) to measure the thermodynamic release of counterions and this term is also corrected in a similar way by the $\Delta H_{ITC}/\Delta H_{DSC}$ factor.

Benzo[a]pyrene Diol Epoxide (BPDE) Adducts

BPDE molecules are metabolized *in vivo* and bind covalently to DNA (*31-32*). Their biological activities are significantly dependent on their stereochemical configurations. Two BPDE adducts have been studied extensively, each is attached to the central dG base of the top strand (Figure 5), using procedures reported earlier (*20, 33*). For instance, (+)-BPDE is strongly tumorigenic whereas its enantiomer (-)-BPDE is not (*34*). The nuclear magnetic resonance (NMR) structures indicate that these adducts lie in the minor groove of the duplex with different orientations: the (+)-anti-BPDE toward the 5' end of the unmodified strand and the (-)-BPDE toward the 3' end of the unmodified strand (*35*).

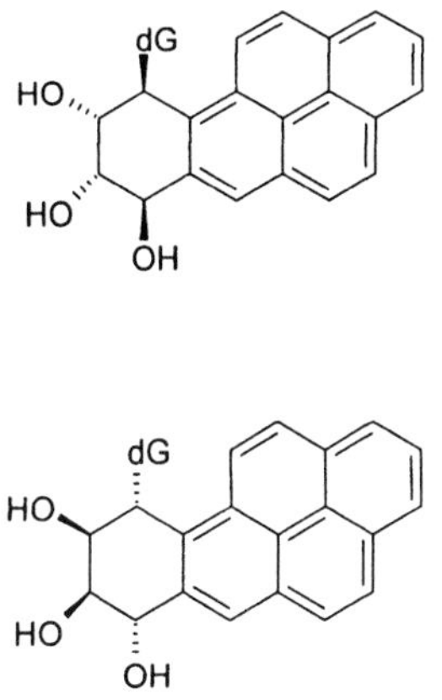

C–C–A–T–C–G–C–T–A–C–C
• • • • • • • • • • • •
G–G–T–A–G–C–G–A–T–G–G
Unmodified duplex

(+) BPDE
C–C–A–T–C–G–C–T–A–C–C
• • • • • • • • • • • •
G–G–T–A–G–C–G–A–T–G–G
(+)-*BPDE* duplex

(–) BPDE
C–C–A–T–C–G–C–T–A–C–C
• • • • • • • • • • • •
G–G–T–A–G–C–G–A–T–G–G
(-)-*BPDE* duplex

Figure 5. BPDE enantiomers (left) and oligonucleotide sequences (right).

The sequence of this duplex (Figure 5) is symmetric with respect to the central modified dG•dC base pair; therefore, the (+)-BPDE is exposing more organic groups to the solvent. In addition, end-to-end ligation of the (+)-BPDE duplexes yields bent structures as seen from their retarded electrophoretic mobility (36).

Complete thermodynamic profiles at 20 °C are shown in Table 3 (20,30). The favorable formation of each duplex, shown in Fig. 5, results from the typical compensation of favorable enthalpy and unfavorable entropy contributions, an uptake of water molecules as seen in the negative ΔV values of Table 3, and the uptake of counterions. The main difference is that the uptake of water for the (+)-BPDE duplex is about 50% larger than for the Unmodified duplex or the (-)-BPDE duplex. The favorable enthalpy term is due to base pairing and base-pair stacking contributions while the unfavorable entropy term results from the increased ordering of two single strands forming a bimolecular complex and immobilization of both water and counterions.

Table 3. Thermodynamic Parameters of Duplex Formation at 20 °C.[a]

Duplex	T_M	ΔH_{ITC}	$\Delta G°_{(20)}$	$T\Delta S_{(20)}$	ΔV	Δn_{Na^+}
	(°C)	(kcal)	(kcal)	(kcal)	(ml)	(mol Na⁺)
Unmodified	62.6	-82	-10.4	-72	-144	-2.90
(+)-BPDE	54.4	-49	-5.1	-44	-209	-2.10
(-)-BPDE	57.7	-47	-5.4	-42	-136	-2.00

[a]All parameters are calculated per mole of duplex and are measured in 20 mM sodium phosphate buffer, 0.1 M NaCl at pH 7.0. The experimental uncertainties are as follow: T_M (±0.5 °C), ΔH_{ITC} (±3%), $\Delta G°$ (±4%), $T\Delta S$ (±5%), ΔV (±7%), and Δn_{Na^+} (±10%).

Relative to the unmodified duplex, the modified duplexes are less stable, by a $\Delta\Delta G°_{(20)}$ of 5.2 kcal/mol due to a lower enthalpy contribution of ~34 kcal/mol, lower entropy term of ~29 kcal/mol. and lower uptake of counterions, $\Delta\Delta n_{Na^+} =$ 0.85 mol Na⁺/mol duplex. Furthermore, we obtained a $\Delta\Delta V$ of -65 ml/mol of duplex for the (+)-BPDE duplex and $\Delta\Delta V$ of +8 ml/mol of duplex for the (-)-BPDE duplex. The comparison of the sign of $\Delta\Delta V$ with the sign of any of the standard thermodynamic functions for a given modified duplex shows that these signs are opposite for the (+)-BPDE duplex while similar signs are obtained for the (-)-BPDE duplex. Therefore, this sign comparison suggests that the formation of the (+)-BPDE duplex experiences a high ordering of structural water while the (-)-BPDE duplex orders a small amount of electrostricted water. In spite that both adducts lie in the minor groove with opposite orientations and reducing base pair stacking, the incorporation of the (+)-BPDE into a DNA duplex yields a higher exposure of hydrophobic surface to the solvent because it immobilizes structural water. The overall effects are similar to the incorporation of bulges into DNA duplexes (*37*). We conclude then that the bent (+)-BPDE duplex has a higher ordering of structural (hydrophobic) water.

Incorporation of Hydrolyzed Cisplatin into a DNA Duplex

The anticancer activity of cisplatin arises from its ability to bind covalently to DNA forming primarily intrastrand crosslinks to adjacent purine residues; the most common adducts are *GG* or *AG* intrastrand cross-links. The incorporation of platinum adducts in a B-DNA helix induces local distortions, causing bending and unwinding of the DNA (*38,39*). The intrastrand crosslink with two guanines, *cis*-[Pt(NH$_3$)$_2${d(GpG-N7(1),N7(2))}], have been characterized extensively by a variety of experimental techniques (*40-44*). The solution structure of the *cis*-[Pt(NH$_3$)$_2${d(GpG-N7(1),N7(2))}] adduct in a dodecamer duplex has been determined by NMR (*40*). The presence of this platinated adduct causes the adjacent guanine bases to roll towards each other by 49°, yielding a helix with an overall helical bend angle of 78° in which the minor groove opposite the adduct site is widened and flattened (*40*). It has been postulated that high-mobility group proteins are involved in the mechanism of cytotoxicity of cisplatin; these proteins recognize and bind selectively to the bent DNA structure and may prevent the action of the excision repair machinery (*45,46*). It has been shown that the presence of this platinated crosslink in an oligomer DNA duplex also changes the overall thermodynamic contributions of DNA (*42-44*). These adduct duplexes are thermodynamically less stable and their lower folding free energies are due to a loss of favorable base-pair stacking interactions (*42-44*), which is influenced by the nature of the bases flanking the lesion (*43*).

Our laboratory has investigated the unfolding thermodynamics of a set of DNA decamer duplexes, *cis*-[Pt(NH$_3$)$_2${d(GpG}] crosslink and unmodified duplex (Figure 6) (*44,47*). Initially, we obtained standard thermodynamic profiles for the unfolding of each duplex using a combination of UV and DSC melting techniques follow up by the investigation of their folding thermodynamics, including hydration effects, with a combination of isothermal titration calorimetry and density techniques at 20 °C. The set of complete thermodynamic profiles at 20 °C are obtained as follows. Unfolded thermodynamic profiles revealed that the *Pt-GG* duplex unfolds with a lower T_M

(and $\Delta G°$) due to lower enthalpy contributions. Folding thermodynamic profiles are shown in Table 4 (*44,47*).

$$H_2N \quad NH_2$$
$$\searrow Pt \nearrow$$
$$\diagup \diagdown$$

```
C-T-C-T-G-G-T-C-T-C          C-T-C-T-G-G-T-C-T-C
• • • • • • • • • •          • • • • • • • • • •
G-A-G-A-C-C-A-G-A-G          G-A-G-A-C-C-A-G-A-G
```

Control   *Pt-GG*

Figure 6. Oligonucleotide sequences, Pt^{2+}-$(NH_3)_2$ is attached to the two central guanine (italicized) at the top of the decamer duplex at the right.

Inspection of these profiles indicates that the favorable formation of each duplex, from mixing its complementary strands, results from the characteristic compensation of a favorable heat and unfavorable entropy contributions. The inclusion of a platinated adduct lowers the heat, by 19.2 kcal/mol, due to unstacking of base pairs. The formation of the *Control* duplex releases water molecules while the *Pt-GG* duplex uptakes water and both duplexes have similar uptakes of counterions.

Table 4. Thermodynamic Profiles of Duplex Formation at 20 °C.[a]

Duplex	T_M, (°C)	ΔH_{ITC}[b] (kca/)	$\Delta G°_{(20)}$ (kca/)	$T\Delta S$ (kcal)	ΔV (ml)	Δn_{Na^+} (mol Na$^+$)
Control	50.3	-65.8	-6.2	-59.6	19	-2.10
Pt-GG	36.8	-46.6	-2.5	-44.1	-35	-2.23

[a] All parameters are calculated per mole of duplex and are measured in 10 mM sodium Hepes buffer at pH 7.5 and 0.1 M NaCl. [b]These heats were measured at 10 °C to insure complete formation of the *Pt-GG* duplex. The experimental uncertainties are as follow: T_M (±0.5 °C), ΔH_{ITC} (±3%), $\Delta G°$ (±4%), $T\Delta S$ (±5%), ΔV (±10%) and Δn_{Na^+} (±5%).

To show that the uptake of water molecules by the *Pt-GG* duplex is structural water, a hypothetical thermodynamic cycle is established using the thermodynamic parameters of the control duplex as a reference state, this assumes that the single strands are considered equivalent. This procedure yielded the following differential thermodynamic profiles: $\Delta\Delta G° = 3.7$ kcal/mol, $\Delta\Delta H_{ITC} = 19.2$ kcal/mol and $\Delta(T\Delta S) = 15.5$ kcal/mol. The positive differential free energy results from a positive compensation of a differential enthalpy term with a differential entropy term. The opposite signs of these parameters with the negative $\Delta\Delta V$ of -54 ml/mol confirm the uptake of structural water by the *Pt-GG* duplex (*30*). Therefore, the incorporation of a platinum adduct in duplex DNA disrupts favorable base-pair stacking interactions, yielding a greater exposure of aromatic bases to the solvent, which in turn immobilizes structural water. The overall effects are consistent with the local DNA structural perturbations induced by the platinum crosslink, which yielded a hydrophobic notch at the lesion site.

In order to understand how the incorporation of a platinated adduct reduces base-pair stacking interactions of duplex DNA and why there is a net uptake of structural water in DNA, the solvent accessible surface area (SASA) (*48*) was calculated for the structures of two dodecamer duplexes. The structure of the platinated duplex is the one determined in solution by NMR (*41*), while the structure of the unplatinated DNA dodecamer in a canonical B-form is generated using the program NAMOT (*49,50*). The overall polar and non-polar surface area of each duplex was modeled with NACCESS (*51,52*). The DNA dodecamer used in the NMR studies has a similar sequence around the lesion site as the decamer duplex of this work. The SASA for the platinated and unmodified dodecamer structures were 4718 Å^2 and 4215 Å^2, respectively. This shows that the platinated duplex has aproximately 503 Å^2 more exposed surface area, polar and nonpolar, than the unmodified duplex. Therefore, the structure of the platinated duplex with a bent conformation is consistent with the conclusion of the thermodynamic studies that the modified DNA decamer is associated with more structural water.

Incorporation of Aminopropyl Cationic Chains into the Dickerson-Drew Dodecamer

To answer the question of how cationic charge can affect DNA structure, ω-aminoalkyl side-chains were appended to the 5-position of deoxypyrimidines and the modified bases incorporated into DNA. It was assumed that these basic chains would salt bridge with the 5'-nonbridging oxygen, neutralizing the remaining unscreened charge (*53,54*), stabilizing DNA duplexes (*55,56*). Gel mobility studies in conjunction with phasing experiments of an A-tract demonstrated that DNA bending occurred. The degree of bending was relative low and dropped from 8° to 4° when the amino group was tethered to the 5-position of dU with a hexyl rather than a propyl chain (*53,54*).

Additional evidence for the ability of ω-aminoalkyl side-chains to bend DNA comes from recent NMR studies of a modified Dickerson-Drew duplex, 5'-d(CGCGAAT$\underline{N}$C^{9}G^{10}CG)-3', where N represents 5-(3-aminopropyl)-

2'deoxyuridine (*57*). The NMR analysis indicates a classical B-DNA structure with normal Watson-Crick base pairing interactions: no significant deviations for ^{1}H chemical shifts were observed with the 3-aminopropyl sidechain (*57*). The structure of *N* and its ability to bend DNA comes from this NMR study that showed the 3-aminopropyl moiety oriented, as predicted, in the 3'-direction from the site of modification (*57*). The presence of the charged amino group in the major groove resulted in a 0.24-ppm downfield shift of one ^{31}P resonance. The perturbed ^{31}P resonance was at the phosphodiester linkage between nucleotides C^9 and G^{10}. However, MD calculations based on the NMR data and electrostatic footprinting (*55,58*) predict that the ammonium group is proximate to the electronegative center at the O^6-position of G^{10} which requires the modified dodecamer to be bent. The back calculation of NOE data using the bent duplex indicated that the bent structure is consistent with all the available NOE data.

Armed with this structural information, our laboratories have investigated the unfolding thermodynamics of a pair of dodecamer duplexes: the Dickerson-Drew dodecamer, which is used as a control molecule, and its modified dodecamer duplex (Figure 7). This sequence is self-complementary; therefore, the duplex will incorporate two aminopropyl cationic chains.

C-G-C-G-A-A-T-T-C-G-C-G   C-G-C-G-A-A-T-*N*-C-G-C-G

• • • • • • • • • • •   • • • • • • • • • • •

G-C-G-C-T-T-A-A-G-C-G-C   G-C-G-C-*N*-T-A-A-G-C-G-C

"Control"   *"Modified"*

Figure 7. Sequences of oligomer duplexes, N represents 5-(3-aminopropyl)-2'deoxyuridine that modifies the dodecamer on the right.

We used a combination of UV melting and DSC techniques to determine standard thermodynamic profiles of the helix-coil transition of each duplex. The ΔH_{DSC} and ΔS_{DSC} parameters are obtained directly from the DSC scans by integration of the experimental ΔC_p vs T and $\Delta C_p/T$ vs T curves, respectively. The $\Delta G°_{(20)}$ is then calculated from the Gibbs equation: $\Delta G°_{(20)} = \Delta H_{DSC} - T\Delta S_{DSC}$. We used UV melting techniques (in conjunction with DSC) as a function of salt concentration to measure Δn_{Na^+}. Unfortunately, we were unable to measure volume changes with these oligonucleotides because the sequences are self-complementary. What we did instead is to use the osmotic stress technique (*59*) to measure the release of water molecules, Δn_W, upon unfolding i.e. UV melting curves were obtained as a function of osmolyte (ethylene glycol) concentration.

UV and DSC melting experiments reveal that each duplex unfolds in biphasic transitions, which corresponds to the duplex-hairpin and hairpin-random coil transitions. However, in this section, we have measured the thermodynamic profiles for the unfolding of each duplex to random coils and report the corresponding folding profiles assuming no changes in the heat capacity between the duplex and random coil states. Complete thermodynamic parameters of duplex formation at 20 °C are shown in Table 5. In low-salt buffer, each duplex fold with a favorable heat and unfavorable entropy contributions, uptake of counterions and uptake of water molecules.

Table 5. Thermodynamic Parameters of Duplex Formation at 20 °C.[a]

Duplex	T_M (°C)	ΔH_{DSC} (kcal)	$\Delta G°_{(20)}$ (kcal)	$T\Delta S_{DSC}$ (kcal)	Δn_{Na^+} (mol Na^+)	Δn_W (mol H_2O)
Control	33.3	-116	-5.0	-111	-2.6	-37
Modified	29.8	-68	-2.2	-66	-1.5	-7

[a] All parameters are calculated per mole of duplex and are measured in 10 mM sodium phosphate buffer at pH 7.0. The experimental uncertainties are as follow: T_M (±0.5 °C), ΔH_{DSC} (±3%), $\Delta G°$ (±4%), $T\Delta S$ (±5%), Δn_{Na^+} (±5%) and Δn_W (±8%).

The thermodynamic contribution of the incorporation of cationic chains in DNA is best illustrated by using a thermodynamic cycle in which the formation of the unmodified duplex is used as the reference reaction. This assumes that the random coils are similar at high temperatures. Relative to the unmodified dodecamer, we obtained an unfavorable $\Delta\Delta G°$ of 2.8 kcal/mol resulting from the compensation of an unfavorable $\Delta\Delta H_{DSC}$ of 48 kcal/mol with a favorable $\Delta(T\Delta S_{DSC})$ term of 45 kcal/mol. The $\Delta\Delta H_{DSC}$ term indicates base-stacking and hydration differences between the two duplexes, assuming that the conformation of the single strands is identical at high temperatures; therefore, the incorporation of a single aminopropyl chain correspond to a 22.5 kcal/mol loss of stacking interactions. The $\Delta(T\Delta S_{DSC})$ term is consistent with both a net counterion release of 1.1 mol Na^+/mol duplex and a net water release of 30 mol water/mol duplex. These effects are consistent with the presence of the tethered amino charge that directly or indirectly neutralizes charge on the phosphate backbone, allowing closer proximity of the aromatic bases. However, the incorporation of a single cationic charge causes a destabilization of the DNA, $\Delta\Delta G° = 1.4$ kcal/mol, which is contrary to what has been observed earlier in similar incorporations at low salt concentration *(60,61)*. To explain this counter-intuitive destabilizing effect, we invoke differences on the neighboring sequences around the modified base and their intrinsic hydration differences.

The decrease of base-pair stacking contributions in the modified duplex (lower folding enthalpy) suggests that the incorporation of an aminopropyl chain in a DNA duplex induces a higher exposure of polar and non-polar groups to the solvent, which is due to a DNA deformation induced by the presence of the cationic side chain that partially neutralizes negative charges generating a small and local helical bend, as has been indicated by the NMR studies *(57)*.

For the interpretation of the lower ion uptake by the modified duplex, it is suggested that the presence of the cationic charged chain in the major groove of this duplex causes a reorganization of the groove counterions around the tethered amine group. The cationic chain would displace counterions and water molecules and partially neutralize negative charges. The combined effects would yield a net unscreening of the backbone phosphates. The exact mechanism responsible for the bending induced by cationic charge associated with the major groove remains to be determined. A reasonable model that accounts for the effect of equilibrium bound, tethered, or covalently attached cations has been presented *(62)*. In this model, cationic charge localized in the major groove will repulse cations *(62)*. The result associated with the phosphate backbone is "naked" phosphates that collapse onto the major groove localized cation (63). This mode of induced bending is distinct from the phosphate neutralization model; however, this latter model could also account for a partially neutralization of negative charge by the cationic aminopropyl chain.

To understand how the aminopropyl induced bending of the DNA would affect the hydrophobic surface exposed to solvent, the SASA of the NMR structures (linear and bent) were determined. The surface areas for the bent (constrained) and linear (unconstrained) NMR structures were 3631 and 3587 Å^2, respectively. This calculates to an increase of 25 Å^2 per modified residue since there are two cationic sidechains in the self-complementary duplex. The same type of surface analysis was done using the same DNA structures except the aminopropyl sidechain was replaced by a hydrogen atom to eliminate any effect due to the location of the sidechain in the major groove. In this case, the bent structure has approximately 33 Å^2 more surface area per modified residue than the linear structure. Therefore, the bent structure from NMR is consistent with the conclusion from the thermodynamic studies that the modified DNA is exposing more polar and non-polar surface to the solvent.

For the interpretation of the lower water uptake by the modified duplex, we suggest that the osmotic stress technique measures the number of water molecules that are tightly associated with DNA. This type of water could very well be the electrostricted water that is associated with the negative phosphate groups of DNA. Alternatively, melting experiments are performed by changing the temperature and these temperature changes may well release gradually structural water, which is immobilized around polar and non-polar groups of the exposed aromatic bases. Direct measurements of the overall hydration of these two duplexes at low temperatures are currently perform using ultrasonic techniques, and will be included in a future report.

CONCLUSIONS

In this review, we focus on the overall thermodynamics of the folding (or unfolding) of DNA oligomer structures that are distorted by the inclusion of small molecules. We report complete thermodynamic profiles, including volume changes, for the physical and chemical placement, of chemical moieties into DNA oligonucleotide duplexes. We were able to shed light on specific contributions to DNA bending, by setting up appropriate thermodynamic cycles i.e., using the unmodified duplexes as control duplexes. Four different systems were discussed: decamer duplexes containing dA•dT tracts; the covalent placement of a benzopyrene derivatives and hydrolyzed cisplatin into duplex DNA; and the inclusion of aminopropyl cationic chains into the Dickerson-Drew dodecamer. In the first three systems direct measurements of the volume change were performed while for the fourth system the volume effects were obtained indirectly by the osmotic stress method. The presence of a bend in all systems has been confirmed indirectly by gel mobility studies or directly by NMR and X-ray crystallographic investigations.

We conclude from the results of the first three systems investigated that the presence of a bend in a DNA duplex is accompanied by a loss of favorable stacking interactions yielding a greater exposure of polar and non-polar groups (or aromatic bases) to the solvent, which in turn are stabilized by the concomitant immobilization of structural water. The overall effect correlates qualitatively with calculations of the solvent accessible surface areas.

Acknowlegments. This work was supported by NIH Grants GM42223 and CA76049.

References

1. Hagerman, P. J. Annu. Rev. Biophys. Biophys. Chem. **1988**, 17, 265.
2. Levitt, M. Proc. Nat. Acad. Sci. USA **1978**, 75, 640.
3. Zhurkin, V. B.; Poltev, V. I.; Florentev, V. L. J. Mol. Biol. **1980**, 14, 882.
4. Hagerman, K. R.; Hagerman, P. J. J. Mol. Biol. **1996**, 260, 207.
5. Baase, W. A.; Schellman, J. A. Makromol. Chem. Macromol. Symp. **1986**, 1, 51.

206

6. Protein Database: http://www.rcsb.org/pdb/
7. Luger, K.; Mader, A. W.; Richmond, R. K.; Sargent, D. F.; Richmond, T. J. Nature **1997**, 389, 251.
8. Westhof, E. Annu. Rev. Bioph. Bioph. Chem. **1998**, 17, 125
9. Saenger, W. *In Principles of Nucleic Acid Structure*; Kantor, G. R., Ed. Springer-Verlag: NewYork, **1984**.
10. Buckin, V. A.; Kankiya, B. I.; Bulichov, N. V.; Lebedev, A. V.; Gukovsky, V. P.; Sarvazyan, A. P.; Williams, A. R. Nature **1989**, 340, 321.
11. Marky, L. A.; Kupke, D. W.;. Biochemistry **1989**, 28, 9982.
12. Rentzeperis, D.; Kupke, D. W.; Marky, L. A. Biopolymers **1993**, 33, 117.
13. Frank, H. S.; Evans, M. W. J. Phys. Chem. **1945**, 13, 507.
14. Kauzman, W. Adv. Protein. Chem. **1959**, 14, 1.
15. Millero, F. J. Chem. Rev. **1971**, 71, 147.
16. Chapman, R. E.; Sturtevant, J. M. Biopolymers **1969**, 7,527.
17. Zieba, K.; Chu, T. M.; Kupke, D. W.; Marky, L. A. Biochemistry **1991**, 30, 8018.
18. Kankia, B. I.; Marky, L. A. J. Phys. Chem. B **1999**, 103, 8759.
19. Marky, L. A.; Kupke, W. Methods Enzymol. **2000**, 323,419.
20. Marky. L. A.; Rentzeperis, D.; Luneva, N. P.; Cosman, M.; Geacintov, N. E.; Kupke, D. W. J. Am. Chem. Soc. **1996**, 118, 3804.
21. Chen, J. H.; Seeman, N. C.; Kallenbach, N. R. Nucleic Acids Res. **1988**, 16, 6803.
22. Hagerman, P. J. Nature **1986**, 32, 449.
23. Sarma, M. H.; Gupta, G.; Garcia, A. E.; Umemoto, K.; Sarma, R. H. Biochemistry. **1990**, 29, 4723.
24. Park, Y. W.; Breslauer, K. J. Proc. Natl. Acad. Sci. USA **1991**, 88, 1551.
25. Alessi, K. Ph.D. Dissertation, New York University, **1995**.
26. SantaLucia, J.; Allawi, H. T.; Seneviratne, P. A. Biochemistry **1996**, 35, 3555.
27. Breslauer, K. J.; Frank, R.; Blocker, H.; Marky, L. A. Proc. Natl. Acad. Sci. USA **1986**, 83, 3746.
28. Krakauer, H. Biopolymers **1972**, 11, 811.
29. Barcelo, F.; Muzard J. P.; Mendoza, R.; Revet, B.; Roques, B. P.; LePecq, J. B. Biochemistry **1991**, 30, 4863.
30. Marky, L. A.; Kupke, D. W.; Kankia, B. I. Methods Enzymol. **2001**, 340, 149.
31. Singer, B.; Grunberger, D. In Molecular Biology of Mutagens and Carcinogens, Plenum Press, New York, **1983**.
32. Singer, B.; Essigman, J. M. Carcinogenesis **1991**, 12, 949.
33. Cosman, M.; Geacintov, N. E.; Harvey, R. G. Carcinogenesis **1990**, 11,1667.

34. Wood, A. W.; Chang, R. L.; Levin, W.; Yagi, H.; Thakker, D. T.; Jerina, D. M.; Conney, A. H. Biochem. Biophys. Res. Com. **1977**, 77, 1389.
35. De los Santos, C.; Cosman, M. ; Hingerty, B. E.; Ibanez, V.; Margulis, L. A.; Geacintov, N. E.; Broyde, S.; Patel, D. J. Biochemistry **1992**, 31, 5245.
36. Mao, B. Ph.D. Dissertation, New York University, **1994**.
37. Zieba, K.; Chu, T. M.; Kupke, D. W.; Marky, L. A. Biochemistry **1991**, 30, 8018.
38. Reedijk, J.; Fichtinger-Schepman, A. M. J.; van Oosterom, A. T.; van de Putte, P. Struct. Bond Berlin **1987**, 67, 53.
39. Fichtinger-Schepman, A. M. J.; van der Veer, J. L.; den Hartog, J. H. J.; Lohman, P. H. M.;Reedijk, J. J. Biochemistry **1985**, 24, 707.
40. Gelasco, A.; Lippard, S. J. Biochemistry **1998**, 37, 9230.
41. Takahara, P. M.; Rosenzweig, A. C.; Frederick, C. A.; Lippard, S. J. Nature **1995**, 377, 649.
42. Poklar, N.' Pilch, D. S.; Lippard, S. J.; Redding, E. A.; Dunham, S. U.; Breslauer, K. J. Proc Natl. Acad. Sci. USA **1996**, 93, 7606.
43. Pilch, D. S.; Dunham, S. U.; Jamieson, E. R.; Lippard, S. J.; Breslauer, K. J. J. Mol. Biol. **2000**, 296, 803.
44. Kankia, B. I.; Kupke, D. W.; Marky, L. A. J. Phys. Chem. B **2001**, 105, 11602.
45. Huang, J. C.; Zamble, D. B.; Reardon, J. T.; Lippard, S. J.; Sancar, A. Proc. Natl. Acad. Sci. USA **1994**, 91, 10394.
46. Zamble, D. B.; Mu, D.; Reardon, J. T. Sancar, A.; Lippard, S. J. Biochemistry **1996**, 35, 10004.
47. Kankia, B. I.; Soto, A. M.; Burns, N.; Shikiya, R.; Tung, C-S.; Marky, L. A. Biopolymers **2002**, 65, 218.
48. Connolly, M. L Science **1983**, 221, 709.
49. Tung, C. S.; Carter, E. S. I. Nucleic Acid MOdeling Tool (NAMOT): An interactive graphic tool for modeling nucleic acid structure; CABIOS, **1994**.
50. Carter, E. S.; Tung, C. S. NAMOT2-A Redesigned Nucleic Acid Modeling Tool: Construction of Non-Canonical DNA Structures; CABIOS, 1996.
51. Hubbard, S. J.; Thornton, J. M. NACCESS; 2.1 ed.; Hubbard, S. J.; Thornton, J. M., Ed.; Department of Biochemistry and Molecular Biology, University College: London.
52. Richards, F. M. Annu. Rev. Biophys. Bioeng. **1977**, 6, 151-176.
53. Strauss, J. K.; Roberts, C.; Nelson, M. G.; Switzer, C.; Maher, L. J. III. Proc. Natl. Acad. Sci. USA **1996**, 93, 9515.
54. Strauss, J. K.; Prakash, T. P.; Roberts, C.; Switzer, C.; Maher, L. J. III. Chem. Biol. **1996**, 3, 671.
55. Dande, P.; Liang, G.; Chen, F. X.; Roberts, C.; Nelson, M. G.; Hashimoto, H.; Switzer, C.; Gold, B. Biochemistry **1997**, 36, 6024.

56. Hashimoto, H.; Nelson, M. G.; Switzer, C. J. Am. Chem. Soc. **1993**, 115, 7128.
57. Li, Z.; Huang, L. ; Dande, P.; Gold, B.; Stone, M. P. J. Am. Chem. Soc. **2002**, 124, 8553.
58. Liang, G.; Encell, L.; Switzer, C.; Gold, B. J. Am. Chem. Soc. **1995**, 117, 10135.
59. Spink, C. H.; Chaires, J. B. Biochemistry **1999**, 38, 496.
60. Soto, A. M.; Kankia, B.; Dande, P.; Gold, B.; Marky, L. A. Nucleic Acids Res. **2001**, 29, 3638.
61. Soto, A. M.; Kankia, B.; Dande, P.; Gold, B.; Marky, L. A. Nucleic Acids Res. **2002**, 30, 3171.
62. Rouzina, I.; Bloomfield, V. A. Biophys. J. **1998**, 74, 3152.
63. Williams, L. D.; Maher, L.J. III Ann. Bioph. Biom. Struc. **2000**, 29, 497.

Chapter 11

Measuring Protein-Induced DNA Bending by Cyclization of Short DNA Fragments Containing Single-Stranded Region

Alexander Vologodskii

Department of Chemistry, New York University, 31 Washington Place, New York, NY 10003

The efficiency of cyclization of short DNA is extremely sensitive to intrinsic or induced bends of the double helix and has been used to measure conformational properties of DNA and DNA-protein complexes. However, DNA rigidity can prevent cyclization in the undisturbed complexes. Also, cyclization of a short DNA fragments, about 200 bp in length, requires proper accounting for the torsional orientation of its ends, which necessitates series of laborious measurements. To overcome these problems we are developing a modification of the method based on cyclization of DNA fragment containing a single-stranded gap near one of the fragment ends. The gap serves as a hinge for lateral and torsional deflections of the double helix. The hinge has to eliminate oscillations of the cyclization efficiency with DNA length that would greatly simplify the measurements and their analysis.

Introduction

In many cases formation of DNA-proteins complexes is associated with strong bending of the double helix. These bends can be very important for subsequent enzymatic transformations of DNA molecules. Thus, it is very useful to have a simple and quantitative method to measure these bends in solution. Sometime FRET can be used for such problems, although it has certain limitations (see (*1*). In particularly, it does not allow to measure distances larger than 10 nm, so its mainly suitable for bends close to 180°. The majority of data on the protein-induced DNA bending in solution was obtained from analysis of DNA cyclization experiments (*2-7*). The method developed by Kahn and Crothers allows one to determine not only the bend angle, but also the torsional deformation of the binding site and corresponding changes of bending and torsional rigidities caused by a protein binding. It requires, however, a large volume of the experimental work and computer optimization to obtain the data. Another approach, based on the mutimerization-cyclization of the oligonucletides 20-40 bp in length (*6, 7*), is simpler, but gives only semiquantitative estimation of the bend angle (*8*). Here we suggest a new approach to the problem, based on the measuring of the cyclization efficiency of DNA fragments about 200 bp in length with a single-stranded gap. Although the approach will allow to measure the bend angle only, it is less laborious and requires simpler theoretical analysis than the method developed by Kahn and Crothers. On the hand, the approach has to allow the measurements of the bent angle with very high accuracy.

Study of the cyclization efficiency has been used for more than twenty years to study DNA internal properties, and has led to unique information about the double helix. It gave extremely elegant proof of the helical nature of the double-stranded DNA (*9*) as well as estimates of the DNA helical repeat and its bending and torsional rigidities (*9-11*). It allowed to study DNA intrinsic bends (*11, 12*). The approach is very attractive because it provides very accurate method to measure DNA conformational properties for DNA fragment with a particular sequence. It is also important that relatively short fragments, needed for the experiments, can be made with nearly any desired sequence. We recently applied the approach to measure contribution of DNA intrinsic curvature to the measured persistence length of the double helix (*13*).

Cyclization of Short DNA Fragments

To determine the DNA conformational parameters for a particular fragment from the cyclization experiments one has to measured its *j*-factor. *j*-Factor defines the efficiency of the fragment cyclization (*14*). It equals the effective concentration of one end of the chain in the vicinity of the other end in the appropriate angular and torsional orientation. Joining DNA ends, either cohesive or blunt, is a slow process, whose rate is not limited by the rate at which these ends diffuse (*15*). Therefore, the *j*-factor also can be expressed over the ratio of the corresponding kinetic constants of irreversible ligation of DNA ends (*9, 16*). On the other hand, the value of the *j*-factor is completely defined by the conformational parameters of the DNA fragment: the minimum energy conformation of its axis, the average bending rigidity of the fragment, its total equilibrium twist and its torsional rigidity. The value of the *j*-factor can be computed with high accuracy for both homogeneous and sequence-dependent models of the double helix, if the corresponding parameters are known (*8, 17, 18*).

j-Factors of short DNA fragments depend very strongly on their lengths (Figure 1). Such dependence allows to extract the value of DNA persistence

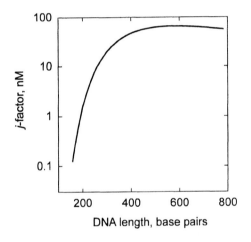

Figure 1. The j*-factors for DNA fragments as a function of their length. The theoretical dependence for the wormlike chain model (19) was converted to the dependence on DNA length assuming that* a = 50 nm. *The dependence does not account for necessity of the proper torsional orientation of the DNA ends and thus represents only axial component of the* j*-factor (*j_l.*in notations of ref. (19)). Reprinted from ref. (13), with permission from Elsevier.*

length for a particular DNA fragment with remarkable accuracy. We calculated, using a theoretical equation for the *j*-factor (*19*), the variation of DNA persistence length as a function of 20% variation *j*-factor. The calculation shows that if we measure the value of j-factor for 200 bp fragments with the relative standard deviation of 20% (what is achievable), we can estimate DNA persistence length from the experimental data with the standard deviation of 2%.

There is one complication in the approach, however. The dependence of the j-factor on the fragment length shown in Figure 1 does not account for the requirement of proper torsional orientation of DNA ends. In its closed circular form the double helix has to make an integer number of turns and this causes extra torsional stress in short DNA circles. As a result of this stress, the oscillations of j-factor with the fragments length are superimposed with the dependence shown in Figure 1 (9). The period of the oscillations corresponds to a DNA helical repeat which is close to 10.5 (20). Its precise value depends on the DNA sequence and thus should be considered as an adjustable parameter during the analysis of experimental data. This means, that measurements of j-factor have to be performed for a few fragment lengths to cover one period of the oscillations. This is laborious, although it also makes determination of the persistence length more reliable and accurate.

To obtain the j-factor experimentally one can measure the ratio of the amounts of circular fragments, $C(t)$, and linear and circular dimers of the fragments, $D(t)$, formed during the early stage of fragment ligation (10):

$$j = 2M_0 \lim_{t \to 0} C(t) / D(t)$$

where M_0 is the initial concentration of the fragments. Since it was shown that the ratio does not change over a wide range of the ligase concentration (10), we can choose it by such away that the time scale of the reaction will be in the convenient range. Circular monomers of the fragments and linear and circular dimers can be separated by gel electrophoresis to measure their relative amounts.

Figure 2 illustrates theoretical fitting of the measured values of j-factors for a set of 11 DNA fragments 196-206 bp in length (13). The theoretical dependence of j-factor on DNA length corresponds to the wormlike chain model (19) and also accounts for the requirement of the proper torsional orientation of the fragment ends (9). The fitting allowed us to determine three conformational parameters of the fragments: the effective persistence length, a, the DNA helical repeat, γ, and the torsional rigidity of the double helix, C. It is important to emphasize that each parameter defines different feature of the curve and thus can be determined unambiguously (10). The best fit corresponds to $a = 49.5 \pm 1$ nm, $\gamma = 10.50 \pm 0.01$ bp/(helix turn), $C = (2.4 \pm 0.1) \cdot 10^{-19}$ erg·cm. To illustrate the accuracy of the persistence length determination from these kind of experimental data we show in Figure 2, in addition to the theoretical curve that provides the best fit to the experimental data, two more curves, for $a = 48.5$ nm and $a = 51.5$ nm. Clearly, these curves do not fit the data.

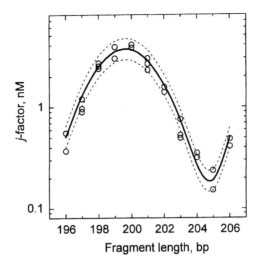

Figure 2. Dependence of j-*factors on the length of intrinsically straight DNA fragments. Experimental data (○) were fitted to the theoretical dependencies by adjusting the values of γ, C and length, a. The best fit (solid line) corresponds to a = 49.5 nm, γ=10.50 bp/(helix turn), C = 2.4·10⁻¹⁹ erg·cm. The dashed lines show theoretical dependencies for a of 48.5 and 50.5 nm and the same values of γ and C. Reprinted from ref. (13), with permission from Elsevier.*

Using this approach we were able to estimate average intrinsic curvature of the double helix (*13*). The measured persistence length of DNA, *a*, depends both on the intrinsic curvature of the double helix and on the thermal fluctuations of the angles between adjacent base pairs (*21-23*). To evaluate two contributions, we determined the value of *a* for two sets of DNA fragments. The first set consisted of DNA fragments 196-206 bp in length with a generic sequence, taken from phage λ and cloned into a plasmid. The second set of 196-206 bp fragments was made up of "intrinsically straight" DNA. DNA sequence of the latter set of fragments consisted of 10 bp segments. In each segment, the sequence of first five bases is repeated by the sequence of the second five bases (Figure 3). Thus, any intrinsic bend in the first half of each segment is compensated by the same bend in the opposite direction in the second half of the segment (assuming that the bends are specified by the sequence of adjacent base pairs).

Actcg**actcg**agcct**agcct**atgac**atgac**acgtt**acgtt**

Figure 3. A fragment of intrinsically straight DNA. Each group of five shadowed bases repeats five preceding bases.

Such intrinsically straight fragments were assembled from synthetic oligonucleotides and cloned into a plasmid. The measured value of a for the intrinsically straight DNA fragments should depend, in a good approximation, on thermal fluctuations only. We found that the values of a for the two types of DNA fragments are very close. Quantitative analysis of this data showed that the contribution of the intrinsic curvature to a is at least 20 times smaller than the contribution of thermal fluctuations. So, if DNA does not contain special elements, first of A-tracts, one can consider that it is intrinsically straight.

Fragments with Single-Stranded Gap

Certainly, the cyclization efficiency of such short fragments is very sensitive to DNA bends as well. However, if we want to apply the method to measure DNA bends, we would find essential limitations. The double helix is very rigid and circular molecules of this size adopt conformations close to a perfect circle. So the ends of a fragment which has a conformation such as shown in the left panel of Figure 4a cannot be joined and ligated without perturbing this conformation. The bend angle has to be reduced to facilitate the cyclization. To overcome this limitation we decided to modify the assay, to make it less perturbing and more accurate. To facilitate cyclization for conformations such as shown in the left panel of Figure 4a, we can incorporate a short single-stranded gap at the end of the fragment.

Figure 4. Conformations of a short DNA fragment in DNA-protein complexes.
(a) A short DNA fragment wrapped around protein. If the bend angle, α,
exceeds 180°, the cyclization perturbs DNA conformation in the complex. (b)
The same DNA fragment with a single-stranded region can be closed without
perturbing the conformation and thus provide more accurate measurement of α.

Single-stranded DNA is very flexible (*24, 25*), so that the gap will serve as a hinge, without substantial restriction of either lateral or torsional deflections (Figure 4b). The single-stranded region adds one more essential technical advantage. It makes all torsional orientations of the ends equivalent, so the cyclization efficiency should no longer depend on the twist of the double-stranded part of the fragment. Thus, it is sufficient to use one fragment length to estimate the value of the bend angle in the fragment.

The DNA fragment for a particular problem can be assembled from two parts. The major part, 160-200 bp in length, will be obtained by cloning and subsequent cutting from a plasmid. A shorter part will consist of three chemically synthesized oligonucleotides. Figure 5 diagrams design of such fragment.

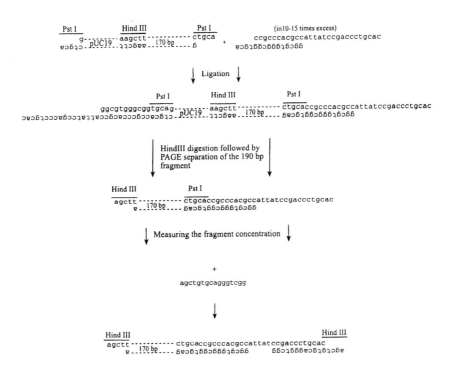

Figure 5. Diagram of assembling of DNA fragment with a single-stranded region. The procedure also includes several steps of cleaning form excess of the short synthetic fragments and enzymes.

Fitting the measured value of the *j*-factor by computed values should allow accurate determination of the bend angle. To illustrate the sensitivity of the method to the value of α, we calculated *j*-factors for a 180 bp DNA fragment bent by the angle α in the middle (Figure 6). One can see that *j*-factor for such fragment changes by more than 4 orders as α increases from $0°$ to $160°$. Since the value of j-factor can be measured with accuracy of about 20%, the method is capable to provide remarkable sensitivity to the value of the bend angle.

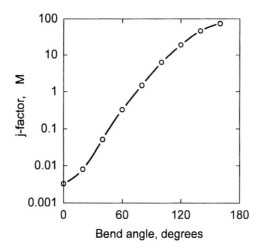

Figure 6. j-*Factor of bent DNA fragment. The value of* j *was computed as a function of the bend angle,* α, *for DNA fragment 180 bp in length. The bend was located in the middle of the segment. It was assumed that in circular form the fragment has a single-stranded gap and DNA segments adjacent to the gap are freely jointed. The computation was performed for the DNA persistence length of 50 nm.*

It is of interest to analyze this sensitivity in some detail. To do so we calculated the distributions between the fragment ends, r, for two values of α (Figure 7). One can see from the figure that the maxima of the distributions correspond to approximately the same values of r. So, methods that are sensitive to the most probable distances, such as FRET, are hardly capable of detecting such a difference. The efficiency of cyclization depends, however, on the probability that $r \approx 0$ rather than on the most probable value of r. The probabilities of $r \approx 0$ for the two values of α differ by a factor of 13.

The value of α for a particular fragment is obtained from comparison of measured j-factors with computed ones, and thus depends on accuracy of the computation. Although the computation can be performed with high accuracy, it assumes a particular model of the DNA-protein complex. We need to make assumptions about the distribution of the bend along the DNA contour in the rigid complex. In many cases, however, we can make reasonable assumptions about the distribution. In other cases we might have to compare different models.

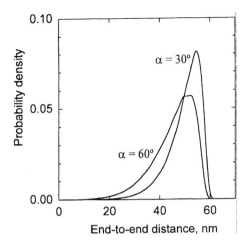

Figure 7. Distribution of the end-to-end distance for bent fragments 180 bp in length. The distributions were computed for α of 30° and 60°.

We believe that this method can become very useful in the studies DNA-protein complexes and DNA conformational properties per se. It is important to note that the method does not require any complex equipment except an imager for quantitative analysis of gels.

We are now refining the procedure to obtain pure fragments with a single-stranded gap. The method places heavy demands on the homogeneity of DNA samples used in the ligation experiment. Indeed only initial small fractions of ligation products are measured so the presence of damaged and potentially more flexible fragments can disturb the results. Thus, the long part of a fragment should always be cloned into a plasmid. Then we will test the method on DNA fragments without proteins that have well defined properties, including proteins with strong intrinsic bends associated with A-tracts.

Acknowledgement

This work was supported by NIH grant GM54215 to the author.

References

1. Hillisch, A.; Lorenz, M.; Diekmann, S. *Curr. Opin. .Struct. Biol.* **2001**, *11*, 201-207.
2. Kahn, J. D.; Crothers, D. M. *Proc. Natl. Acad. Sci. USA* **1992**, *89*, 6343-6347.
3. Kahn, J. D.; Crothers, D. M. *Cold Spring Harbor Symp. Quant. Biol.* **1993**, *58*, 115-122.
4. Kahn, J. D.; Crothers, D. M. *J. Mol. Biol.* **1998**, *276*, 287-309.
5. Davis, N. A.; Majee, S. S.; Kahn, J. D. *J Mol Biol* **1999**, *291*, 249-265.
6. Lyubchenko, Y. L.; Shlyakhtenko, L. S.; Chernov, B.; Harrington, R. E. *Proc. Natl. Acad. Sci. USA* **1991**, *88*, 5331-5334.
7. Balagurumoorthy, P.; Sakamoto, H.; Lewis, M. S.; Zambrano, N.; Clore, G. M.; Gronenborn, A. M.; Appella, E.; Harrington, R. E. *Proc. Natl. Acad. Sci. USA* **1995**, *92*, 8591-8595.
8. Podtelezhnikov, A. A.; Mao, C.; Seeman, N. C.; Vologodskii, A. V. *Biophys. J* **2000**, *79*, 2692-2704.
9. Shore, D.; Baldwin, R. L. *J. Mol. Biol.* **1983**, *170*, 957-981.
10. Taylor, W. H.; Hagerman, P. J. *J. Mol. Biol.* **1990**, *212*, 363-376.
11. Crothers, D. M.; Drak, J.; Kahn, J. D.; Levene, S. D. *Methods in Enzymology* **1992**, *212*, 3-29.
12. Roychoudhury, M.; Sitlani, A.; Lapham, J.; Crothers, D. M. *Proc Natl Acad Sci U S A* **2000**, *97*, 13608-13613.
13. Vologodskaia, M.; Vologodskii, A. *J. Mol. Biol.* **2002**, *317*, 205-213.
14. Jacobson, H.; Stockmayer, W. H. *J. Chem. Phys.* **1950**, *18*, 1600-1606.
15. Wang, J. C.; Davidson, N. *Cold Spring Harbor Symp. Quant. Biol.* **1968**, *33*, 409-415.
16. Shore, D.; Langowski, J.; Baldwin, R. L. *Proc. Natl. Acad. Sci. USA* **1981**, *78*, 4833-4837.
17. Hagerman, P. J. *Annu. Rev. Biochem.* **1990**, *59*, 755-781.
18. Koo, H. S.; Drak, J.; Rice, J. A.; Crothers, D. M. *Biochemistry* **1990**, *29*, 4227-4234.
19. Shimada, J.; Yamakawa, H. *Macromolecules* **1984**, *17*, 689-698.
20. Wang, J. C. *Proc. Natl. Acad. Sci. USA* **1979**, *76*, 200-203.
21. Trifonov, E. N.; Tan, R. K. Z.; Harvey, S. C. In *DNA bending and curvature*; Olson, W. K.; Sarma, M. H.; Sarma, R. H.; Sundaralingam, M., Eds.; Adenine Press: New York, 1988; pp 243-253.
22. Schellman, J. A.; Harvey, S. C. *Biophys. Chem.* **1995**, *55*, 95-114.
23. Katritch, V.; Vologodskii, A. *Biophys. J.* **1997**, *72*, 1070-1079.
24. Mills, J. B.; Cooper, J. P.; Hagerman, P. J. *Biochemistry* **1994**, *33*, 1797-1803.
25. Smith, S. B.; Cui, Y.; Bustamante, C. *Science* **1996**, *271*, 795-799.

Chapter 12

DNA Sequence-Dependent Curvature and Flexibility in Stability and Organization of Nucleosomes

P. De Santis[1], M. Savino[2], A. Scipioni[1], and C. Anselmi[1]

[1]Dipartimento di Chimica and [2]Dipartimento di Genetica e Biologia Molecolare, Università di Roma La Sapienza P.le A. Moro 5, 00185 Roma, Italy

Static and dynamic DNA curvature are involved in fundamental biological functions as well as in the stability of nucleosomes and their organization in the chromatin architecture. We have developed a statistical mechanics model to derive superstructural properties of DNA from the sequence-dependent curvature and flexibility. Very recently this model allows us to predict the unexpected relative instability of reconstituted nucleosomes of a highly curved Crithidia *fasciculata* DNA tract. Attempts are in progress to reproduce nucleosome packaging in large genome tracts as a result of the predicted phasing and translational positioning of the nucleosomes along the sequence.

Static and dynamic curvatures are involved in fundamental DNA functions. Such sequence-dependent DNA superstructural features influence the circularization propensity as well as the writing transitions in topological domains. However, their main role is related to the recognition mechanisms involving DNA binding proteins. The nucleosome is the DNA association complex with the histone octamer and represents the elemental unit of chromatin. Its structure is characterized by a flat solenoid-like structure in which a DNA tract of 146 bp is wrapped around the histone core with a pseudo-dyad symmetry (*1*). The deep knowledge of the molecular structure, which reveals a large mass of details of DNA and protein core, including over 3000 water molecules, provides a dramatic image of the great complexity of the problem about the preferential positioning and stability of the nucleosome along DNA.

Therefore, the question of the sequence determinants of the DNA-histone recognition appears at present to be a complex and a still debated problem. Competitive reconstitution experiments allow the determination of the differential thermodynamic nucleosome affinity along a DNA sequence, providing a sound basis for discovering the sequence effects on nucleosome stability (*2-15*).

The original hypothesis was that intrinsically curved DNA, characterized by phased sequences of AA (TT) dinucleotide steps, could have a large propensity to form nucleosomes (*16-20*). However, from these investigations a complex role of curvature emerged. Shrader and Crothers (*2, 3*) found that some intrinsically curved DNAs, which were supposed to form highly stable nucleosomes, surprisingly showed lower affinity for the histone octamer than relatively straight DNAs with similar sequences. DNAs, isolated in the mouse genome, characterized by runs of three or four adenine phased residues, extensive CA repeats, and TATA tetranucleotides, were found to form very stable nucleosomes despite their low integral curvature (*8*). Further, SELEX experiments, carried out with a large pool of random DNA fragments, allowed the isolation of individuals having the highest affinity with histone octamer so far obtained although they are characterized by relatively low curvature (*10*).

The complexity of the available data and the experimental evidence that the whole range of differential affinity of the histone octamer for DNAs is restricted within a few kcal/nucleosome suggested that the nucleosome stability differences could be due to occasional registers of few amino acid and nucleotide residues. Therefore, an alternative model was proposed which tries to explain the stability of the DNA-histone association as due to few chemical determinants, some authors localize nearby to the nucleosome dyad axis (*16-20*).

Recently, the complexity of the nucleosome crystal structure and the very large number of DNA-histone-water interactions prompted us to consider the differential stability of nucleosomes as a statistical thermodynamic problem where the average properties of the ensemble dominate over the local specific interactions. Therefore, we developed a statistical mechanics model to derive the differential affinity of DNA for the histone octamer from the sequence-dependent curvature and flexibility (*21, 22*).

The model

If $\Delta G(k)$ represents the nucleosome reconstitution free energy difference of the kth DNA tract with $L=146$ bp along a sequence with N bp, the free energy per mole of nucleosome, ΔG, pertinent to the whole DNA is:

$$\beta \Delta G = -ln \sum_{k=L/2}^{N-L/2} \exp\left[-\beta \Delta G(k)\right] \tag{1}$$

where β is $1/RT$. The exponential term represents the equilibrium constant pertinent to the nucleosome reconstitution at the kth DNA tract.

The relation with the pertinent canonical partition functions allows us to write the nucleosome reconstitution free-energy difference as

$$\beta \Delta G(k) = -ln\frac{Q_n(k)}{Q_n^*} + ln\frac{Q_f(k)}{Q_f^*} \tag{2}$$

where $Q_n(k)$ is the configurational canonical partition function of the kth nucleosomal DNA tract along the sequence; $Q_f(k)$ is that relative to the corresponding free DNA, and Q_n^* and Q_f^* are those pertinent to an ideal standard intrinsically straight DNA with a random sequence. The partition functions of the remaining DNA tracts not involved in the kth nucleosome are considered in a first approximation equivalent to those of the free DNA and cancel in the ratio. We evaluated the elastic contributions to the partition functions, related to the sum of the bending and twisting, energies necessary to distort the intrinsic structure of the kth DNA tract in the nucleosomal form, assuming first order elasticity and obtained (21, 22):

$$\beta \Delta G_{el}(k) = \beta \Delta E^{\circ}(k) - 3/2L\ln\left\langle\frac{T}{T^*}\right\rangle + Z - \ln J_0(iZ) \tag{3}$$

where $\Delta E^{\circ}(k)$ is the minimum elastic energy required to distort the L bp kth tract in the nucleosomal form. $\langle T/T^*\rangle$ is the average dinucleotide empirical melting temperature of the kth DNA tract and represents its average relative rigidity with respect to a standard DNA as recently assessed by the analysis of AFM images (31-33) (23-25). Z is equal to $(\beta b/L)\langle T/T^*\rangle A_n A_f^{\circ}$, where A_n and A_f° are the Fourier transform amplitudes of frequency 0.17 of the nucleosome and the free DNA curvature function along the kth tract of the sequence. It is worth noting that $A_n A_f^{\circ}$ represents the correlation between the superstructure of the nucleosomal DNA and that of the free form, according to the convolution theorem. $J_0(iZ)$ is the zero-order Bessel function of the imaginary argument Z.

Alternatively, it is possible to evaluate the free energy pertinent to each nucleosome position along DNA, by constraining its dyad axis to lie on the pseudo-dyad axis in the large groove of recurrent base pairs according to the X-ray structure (1). In this case, the logarithm of the Bessel function in eq 3 is replaced by a periodic function of the angle, ϕ, between the directions of the effective intrinsic curvature and the major groove axis.

$$\beta \, \Delta G_{el}\left(k\right) = \beta \Delta E^{o}\left(k\right) - \frac{3}{2} L \, \ln \left\langle \frac{T}{T^{*}} \right\rangle + Z - Z \cos \varphi \qquad (4)$$

Eq 4 produces an evident tenfold periodicity in the trend of the nucleosome stability along a DNA sequence, especially for curved tracts, in which the bending energy differences between in-phase and out-of-phase nucleosomes are dramatic. As expected, intrinsic curvature in phase with the induced nucleosomal curvature increases the nucleosome stability.

To calculate the functions involved in eqs 3,4 we adopted our set of roll tilt and twist angles for each dinucleotide step as well as the related rigidity factors, T/T^{*}, which modulate the apparent average bending and twisting elastic constants as reported in our previous papers (21, 22).

Results

We evaluated the equilibrium constant of the competitive nucleosome reconstitution in terms of thermodynamic and structural parameters of the dinucleotide steps adopting first-order elasticity to calculate the pertinent canonical partition functions involved in the nucleosome formation. The theoretical free energy values so obtained showed satisfactory agreement with the experimental data for a number of DNAs but major deviations for others. This disagreement, however, was strictly correlated ($R = 0.99$) with the free DNA effective curvature, $\langle A_f \rangle$, which represents in modulus and phase the degree of similarity of the free DNA curvature with that of the nucleosome. This strongly indicated the existence of an additional curvature-dependent contribution to the free energy, which appears to destabilize the nucleosome. Such a contribution was obtained by fitting the free energy deviations by a simple parabolic function of the effective curvature (21, 22). More recently, the functional dependence of such interactions was slightly modified as 5.35 $\langle A_f \rangle^{1.5}$ in RT unit. This does not significantly change the free energy values pertinent to the DNA tracts early calculated, but allows a better formulation of this semi-empirical contribution in a practically whole range of DNA curvatures. This contribution of the DNA intrinsic curvature, which appears to destabilize the nucleosome formation, could be related to any curvature dependent property. The hypothesis of preferential interactions of nucleotide steps with the histones, in competition with those that determine the curvature, does not seem to be supported by the Drew and Travers findings. In fact, the phase of DNA curvature in a reconstituted nucleosome is the same as in the related circularized tract (18).

We interpreted this free energy contribution as due to the groove contractions in intrinsically curved free DNAs, which stabilizes the water spine and counterion interactions adding a further energy cost to the nucleosome formation. A part of these interactions is lost and substituted by the DNA-histone moiety interactions (*26-33*).

Figures 1 and 2 illustrate the comparison between the experimental and theoretical free energy values of about 60 DNA tracts (see Table I) 16 more than those reported in our previous papers (*21, 22*) corresponding to a linear correlation ($R = 0.90$). It is very striking that the nucleosome stability of the well-known highly curved tract of Crithidia *fasciculata* is found being characterized by high free energy (to be published) comparable with that of telomers, a class of relatively less stable nucleosomes (*11*).

This DNA fragment was obtained from Crithidia kinetoplast minicircle that produced the following 223-bp fragment with the original Crithidia 211 curved tract after being inserted in the plasmid *p*PK 201/cat and digestion with BamH1:

GATCCCGCCT	AAAATTCCAA	CCGAAAATCG	CGAGGTTACT
TTTTTGGAGC	CCGAAAACCA	CCCAAAATCA	AGGAAAAATG
GCCAAAAAAT	GCCAAAAAAT	AGCGAAAATA	CCCCGAAAAT
TGGCAAAAAT	TAACAAAAAA	TAGCGAATTT	CCCTGAATTT
TAGGCGAAAA	AACCCCCGAA	AATGGCCAAA	AACGCACTGA
AAATCAAAAT	CTGAACGTCT	CGG	

It is characterized by in phase repetitions of 4-5 AA dinucleotide steps that should face in the histone surface. In spite of all the apparent favorable conditions (high curvature and suitable distribution towards the histone core of the AA steps (*16*)), for a strong affinity with the histone octamer, the Crithidia *fasciculata* DNA shows, in agreement with our model, a low thermodynamic association constant (see Figures 1 and 2). This result further supports the issue that DNA intrinsic curvature plays a dual role by decreasing the distortion energy to assume the nucleosomal shape, but increasing the free-energy cost of releasing a part of the water spine consequent to the nucleosome formation. As a consequence of this dual role of the intrinsic curvature, the existence of a minimum in the free energy profile versus the curvature is expected for similar sized DNA tracts with comparable flexibility. To illustrate this issue, Figure 3 reports the nucleosome competitive reconstitution free energy trend versus the effective curvature of a large pool of DNA tracts characterized by a close range of lengths (160-220 bp). Similar quadratic functions interpolate both the experimental and theoretical data indicating an optimal curvature at 1.5 rad/nucl, corresponding to the Shrader and Crothers TG pentamer nucleosome (*2, 3*) we adopted as a reference of the nucleosome reconstitution free energy.

Organization of nucleosomes in chromatin fiber

The satisfactory results obtained on the nucleosome competitive reconstitution in DNA tracts, generally longer than 146 bp, implicitly support the possibility to localize the nucleosome virtual positioning at the free energy

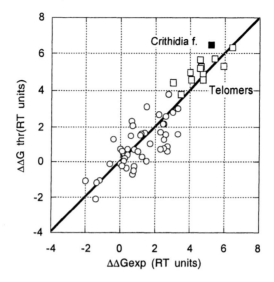

Figure 1. Comparison between theoretical and experimental nucleosome reconstitution free energies (RT units) for the Crithidia fragment (■), the telomers (□) and the other sequences (O). Data (2-5,8-15) are related to the TG pentamer as a standard (2)

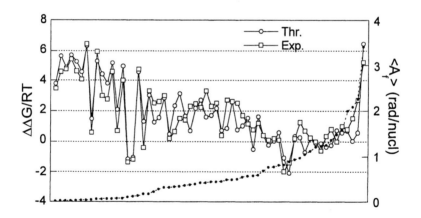

Figure 2. Comparison between experimental and theoretical free nucleosome reconstitution energy for the DNA tracts reported in Table I, sorted at increasing effective curvature $\langle A_f \rangle$

minima along the sequence (eq 4) *(15)*. This prompted us to investigate the packing of nucleosomes in the chromatin fiber.

We assumed that the large-scale organization of chromatin fiber is driven by the optimization of the nucleosome close packing, conditioned by the entry-exit angle of the linkers and their length. Once the entry-exit angle is fixed, changing the linker length determines the mutual orientation of nucleosomes in a discrete and periodical way with the period of the B DNA. As a result, similar structures are obtained changing the linkers of integral numbers of DNA helical turns. Starting from these discrete structures, the packing optimization is obtained by elastic deformations of the linkers adopting standard bending and twisting force constants *(21,22)*.

We approached the problem adopting an array of three nucleosomes as the independent unit of the chromatin fiber. Therefore, we calculated the packing energy in terms of the length of the two linkers in bp, adopting a Gay-Berne potential *(34)*, with a minimum of 3.4 kT at the contact of the ellipsoidal surfaces enveloping nucleosomes (as suggested by the laser tweezers stretching experiments by Cui and Bustamante *(35)*). An example is shown in Figure 4 for the entry-exit linker angle of 35° *(36)*. The continuity of the energy surface is obtained by torsional distortion of the linkers. The diagram covers the linker interval between 40 and 51 bp and practically reproduces periodically for longer or shorter linkers. The periodicity of the packing energy diagram indicates that the relative phasing of the nucleosomes is the main factor of the chromatin architecture. The structures shown in Figure 5, corresponding to the two energy minima, fulfill the experimental features of the 30 nm chromatin fiber as well as the about orthogonality of the nucleosome axis with respect to the fiber axis *(36-38)*. It is interesting that the main interactions are between the even or the odd positioned nucleosomes since the nearest-neighbor interactions are negligible because of the linker bending rigidity.

Concluding remarks

We can advance the conclusion that the intrinsic curvature is the main factor that controls nucleosome stability and consequently, nucleosome positioning. It decreases the distortion energy required to transform the free DNA tract in the nucleosomal shape, but increases the free-energy cost for releasing a part of the water spine, consequent to the nucleosome formation, which is replaced by the interactions with the histones. The flexibility also seems to have a dual role: decreasing the distortion energy for nucleosome formation and increasing the entropy difference between the flexible free DNA and the final rather rigid nucleosome structure as previously shown *(21, 22)*. Therefore the great number and complexity of the interactions between DNA and histones, which involve also a very large number of water molecules, plausibly results in a rather general average force field which appears to be scarcely influenced by the DNA sequence.

Table I. The whole set of DNA fragments ordered by increasing value of curvature, $\langle A_f \rangle$.

Sequence	Reference	$\langle A_f \rangle$ (rad/nucl)	$\Delta\Delta G_{exp}$ (kcal/mol)
B. mori telomer	(14)	0.02	2.70
H. sapiens telomer (254 bp)	(14)	0.02	2.80
H. sapiens telomer (192 bp)	(14)	0.02	3.20
H. sapiens telomer (222 bp)	This paper	0.03	2.90
A. thaliana telomer (236 bp)	(14)	0.03	2.40
T. termophila telomer	(14)	0.04	3.80
CGG_{74}	(5)	0.04	0.34
S. cerevisiae telomer	(14)	0.06	3.50
TRGC	(2)	0.07	1.30
A. thaliana telomer (195 bp)	(14)	0.07	2.75
CGG_{13}	(5)	0.07	0.42
CTG^{62}	(4)	0.10	−0.80
CTG^{55}	(4)	0.12	−0.71
C. reinhardtii telomer	(14)	0.13	2.80
CTG^{10}	(5)	0.16	−0.24
34	(3)	0.17	2.00
L. variegatus	Personal communication	0.22	1.50
20	(3)	0.23	1.55
AOUT	(3)	0.31	1.80
CAG-runs-CAG	(8)	0.32	0.12
NoSecs-1	(8)	0.33	0.40
19	(3)	0.34	0.90
BADSECS-2	(13)	0.36	0.84
X. borealis 5S	Personal communication	0.38	1.35
TGGA-3	(13)	0.39	1.47
IGC	(3)	0.42	1.30
TAND-1	(14, 16)	0.42	2.00
TGGA-1	(13)	0.44	1.38
TGA	(13)	0.44	1.50
5S RNA gene	(5)	0.46	0.25
22	(3)	0.49	1.60
EXGC	(3)	0.49	1.55
TGGA-2	(13)	0.50	1.50

Continued on next page.

Table I. The whole set of DNA fragments ordered by increasing value of curvature, $\langle A_f \rangle$.

Sequence	Reference	$\langle A_f \rangle$ (rad/nucl)	$\Delta\Delta G_{exp}$ (kcal/mol)
BADSECS-1	*(13)*	0.55	1.02
SCEN6	*(12)*	0.57	0.70
EXAT	*(3)*	0.59	0.70
CA-runs-CA-1	*(8)*	0.68	0.30
H4 ΔCTG	*(4)*	0.77	0.04
H4 ΔCTG/CGG	*(4)*	0.78	0.12
Mouse minor satellite	*(8)*	0.81	0.06
Histone H4 gene	*(4)*	0.82	−1.21
618	*(10)*	0.88	−2.10
A-tracts A-1	*(8)*	0.94	0.12
ANISO	*(3)*	0.97	0.75
FIN	*(3)*	1.05	0.42
ANNA	*(3)*	1.13	0.15
TG	*(2)*	1.21	0
TATA-tetrads-TATA	*(8)*	1.24	−0.35
KlCEN1	*(12)*	1.26	0.20
IAT	*(3)*	1.37	0.50
GT	*(2)*	1.49	0.00
TIATR	*(3)*	1.58	0.60
AEXT	*(3)*	1.99	0.50
TTT	*(3)*	2.09	0.90
END	*(3)*	2.26	1.60
Crithidia fragment (223 bp)	This paper	3.42	3.20

228

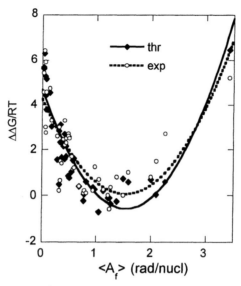

Figure 3. Experimental (◆) and theoretical (O) nucleosome reconstitution free energy versus the effective curvature, ⟨A_f⟩, for a homogeneous subset of DNA fragments (160-220 bp).

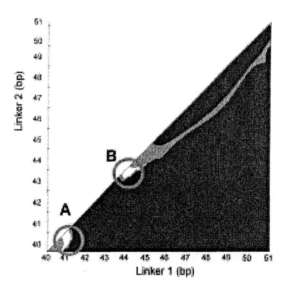

Figure 4. Packing energy diagram of a three-nucleosome unit in terms of the length of the two linkers. The two most compact fiber architectures are indicated. The entry-exit angle is fixed to 35° according to Electron Cryomicroscopy images at high ionic strength (36).

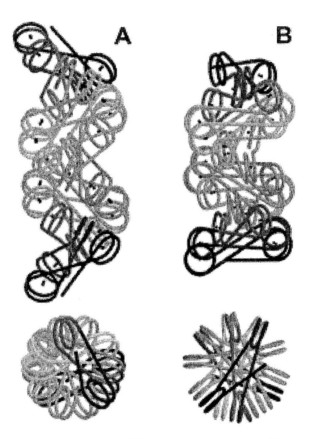

Figure 5. Schematic pictures of the two most compact architectures of chromatin fiber corresponding to the minima A and B of Figure 4

The results of the best packing of nucleosomes in the chromatin fiber indicate the main role of the nucleosome phasing consequent to their relative positioning along DNA. The resulting structures could be interpreted as a sort of a double helix architectures since the main interactions are localized between odd or even nucleosome positions along DNA. This feature permits to investigate the observed winding of chromatin fiber in terms of dinucleosome orientation parameters, using a representation similar to that adopted for the DNA intrinsic curvature in terms of the dinucleotide steps orientation angles (to be published).

Attempts are in progress to find the virtual chromatin architecture of genome tracts starting from the nucleosome positioning obtained by localizing the deepest free energy minima along the sequence.

References

1. Luger, K.; Mäder, A. W.; Richmond, R. K.; Sargent, D. F.; Richmond, T. J. *Nature* **1997**, *389, 251-260*.
2. Shrader, T. E.; Crothers, D. M. *Proc. Natl. Acad. Sci. USA* **1989**, *86*, 7418-7422.
3. Shrader, T. E.; Crothers, D. M. *J. Mol. Biol.* **1990**, *216, 69-84*.
4. Godde, J. S.; Wolffe, A. P. *J. Biol. Chem.* **1996**, *271, 15222-15229*.
5. Godde, J. S.; Kass, S. U.; Hirst, M. C.; Wolffe, A. P. *J. Biol. Chem.* **1996**, *271, 24325-24328*.
6. Wang, Y.; Gellibolian, R.; Shimizu, M.; Wells, R. D.; Griffith, J. *J. Mol. Biol.* **1996**, *263, 511-516*.
7. Wang, Y.; Griffith, J. *Proc. Natl. Acad. Sci. USA* **1996**, *93, 8863-8867*.
8. Widlund, H. R.; Cao, H.; Simonsson, S.; Magnusson, E.; Simonsson, T.; Nielsen, P. E.; Kahn, J. D.; Crothers, D.M.; Kubista, M. *J. Mol. Biol.* **1997**, *267, 807-817*.
9. Cacchione, S.; Cerone, M. A.; Savino, M. *FEBS Lett.* **1997**, *400, 37-41*.
10. Lowary, P. T.; Widom, J. *J. Mol. Biol.* **1998**, *276, 19-42*.
11. Rossetti, L.; Cacchione, S., Fuà, M.; Savino, M. *Biochemistry* **1998**, *37*, 6727-6737.
12. Del Cornò, M.; De Santis, P.; Sampaolese, B.; Savino, M. *FEBS Lett.* **1998**, *431, 66-70*.
13. Cao, H.; Widlund, H. R.; Simonsson, T.; Kubista, M. *J. Mol. Biol.* **1998**, *281, 253-260*.
14. Filesi, I.; Cacchione, S.; De Santis, P.; Rossetti, L.; Savino, M. *Biophys. Chem.* **1999**, *83, 223-237*.

15. Mattei, S.; Sampaolese, B.; De Santis, P.; Savino, M. *Biophys. Chem.* **2002**, *97, 173-187.*

16. Satchwell, S.; Drew, H. R.; Travers, A. A. *J. Mol Biol.* **1986**, *191, 679-675.*

17. McGhee, J. D.; Felsenfeld, G. *Annu. Rev. Biochem*, **1980**, *49,* 1115-1156.

18. Drew, H. R.; Travers, A. A. *J. Mol. Biol.* **1985**, *186, 773-790.*

19. Widom, J. *Annu. Rev. Biophys. Chem.*, **1989**, *18, 365-395.*

20. Blank, T. A.; Becker, P. B. *J. Mol. Biol.* **1996**, *260, 1-8.*

21. Anselmi, C.; Bocchinfuso, G.; De Santis, P.; Savino, M.; Scipioni, A. *J. Mol. Biol.* **1999**, *286, 1293-1301.*

22. Anselmi, C.; Bocchinfuso, G.; De Santis, P.; Savino, M.; Scipioni, A. *Biophys. J.* **2000**, *79, 601-613.*

23. Scipioni, A.; Anselmi, C.; Zuccheri, G.; Samorì, B.; De Santis, P. *Biophys. J.* **2002**, *83, 2408-2418.*

24. Scipioni, A.; Zuccheri, G.; Anselmi; C., Bergia, A.; Samorì, B.; De Santis, P. *Chem. Biol.* **2002**, *9, 1315-1321.*

25. Anselmi, C.; De Santis, P.; Paparcone, R.; Savino, M.; Scipioni, A. *Biophys. Chem.* **2002**, *95, 23-47.*

26. Drew, H. R.; Dickerson, R. E. *J. Mol. Biol.* **1981**, *151, 535-556.*

27. Liepinsh, E.; Otting, G.; Wuthrich, K. *Nucleic Acid Res.* **1992**, *20, 4549-4553.*

28. Berman, H. M. *Curr. Opin. Struct. Biol.* **1994**, *4, 345-350.*

29. Shui, X.; McFail-Isom, L.; Hu, G. G.; Williams, L. D. *Biochemistry* **1998**, *37, 8341-8355.*

30. Shui, X.; Sines, C. C.; McFail-Isom, L.; VanDerveer, D.; William, L. D. *Biochemistry* **1998**, *37, 16877-16887.*

31. McFail-Isom, L.; Sines, C.C.; William, L.D. *Curr. Opin. Struct Biol.* **1999**, *9, 298-308.*

32. Gottesfeld, J.M; Luger K. *Biochemistry* **2001**, *40, 10927-10933.*

33. Hud, N.V.; Sklenar V.; Feigon, J. *J. Mol. Biol.* **1999**, *286, 651-660.*

34. Gay, J. G.; Berne, B. J. *J. Chem. Phys.* **1981**, *74, 3316-3319.*

35. Cui, Y.; Bustamante, C. *Proc. Natl. Acad. Sci. USA* **2000**, *97, 127-132.*

36. Bednar, J.; Horowitz, R. A.; Grigoryev, S. A.; Carruthers L. M.; Hansen, J. F.; Koster, A. J; Woodcock, C. L. *Proc. Natl. Acad. Sci. USA* **1998**, *95, 14173-14178.*

37. van Holde, K. E. In Chromatin;Springer Series in Molecular Biology; Rich, A., Ed.; Springer-Verlag: New York, 1988; pp 289-317.

38. Richmond, T. J.; Widom, J. In Chromatin Structure and Gene Expression; Elgin, S. C. R., Workman, J. L. Eds; Frontiers in Molecular Biology 35, Oxford University Press: Oxford, UK, 2000; pp 1-19.

Indexes

Author Index

Subject Index

A

Accessible surface areas, nuclear magnetic resonance (NMR) structures, 74–75, 78*f*

Aminopropyl cationic chains
incorporation into Dickerson–Drew dodecamer, 201–205
induced bending of DNA, 204
lower ion uptake by modified duplex, 204
lower water uptake by modified duplex, 204–205
thermodynamic parameters of duplex formation, 203*t*

Anticancer activity, cisplatin, 199

Association reactions, volume changes, 191–192

Asymmetric charge neutralization
bend angle, 137–138
bend angle in single line model, 139
change in force after neutralization, 137
comparative gel retardation experiments, 134
DNA four-way junction, 140
driving force for bending, 134
generalization of Manning model, 138–139
generalization of model, 140
potential of mean force approach, 140
predicted bend angles, 141*t*
stretching force in divalent salt solution, 138–139
theoretical methods, 135–138
theory, 136–137
two line charges and in-line bend angles, 139

A-tract bending
models, 34–35
molecular basis, 4

A-tract deoxyribonucleic acid (DNA)
accessible surface areas of A_6 and $A_2G_1A_3$ nuclear magnetic resonance (NMR) structures, 78*f*
AT to TA transversions within A-tract, 82
base inclination within A-tracts, 76*f*, 77*f*
cation binding, 82
change in inclination between C10 and G11 in A_6 NMR structure, 78*f*
characteristics of A-tracts in solution, 66–67
comparison A-tract structures by superposition rms distances, 74*t*
comparison of A-tract DNA structures, 71, 72*f*
direction of curvature, 67, 71
geometry of curvature, 68*f*, 69*f*
mechanism of induced curvature, 71, 74–75, 80
minor groove distances, 75, 79*f*
minor groove hydration within, 81*f*
model of A-tract curvature, 83–85
models of A-tract curvature, 67, 70*f*
NMR, 65–66
normal vector plots of structures, 71, 73*f*
purine-purine stacking within A-tract, 82–83
reasons for A-tract curvature, 80, 82–83
residual dipolar couplings (RDC), 66
role of monovalent cations in DNA curvature, 82
single crystal X-ray diffraction, 65

238

See also Deoxyribonucleic acid
(DNA)
AT to TA transversions, middle of A-
tract, 82

B

Base pair sequence
average structures and calculated
bending angles, 31*f*
calorimetric and spectroscopic
results, 30*f*
effects on DNA structure, 26, 28–29
molecular dynamics calculated
effects, 28–29
sodium ion occupancies per DNA
base pair, 27*f*
Bend angle, evaluating due to charge
neutralization, 137–138
Bending
aminopropyl induced, of DNA, 204
asymmetric charge neutralization,
141*t*
asymmetric enhancement of
phosphate repulsion, 119, 123,
124, 129–130
DNA, 3
DNA axis, 34–36
DNA bending by tethered
phosphonate dianion in major
groove, 127*t*
DNA helix, 6
driving force by asymmetric charge
neutralization, 134
electrostatic theory of phosphate
distribution in bent DNA, 115–
116, 119
hypothetical, by electrostatic
collapse, 120*f*
phosphate crowding and DNA, 112–
113
semi-synthetic phasing analysis of
DNA, 114–115
See also Asymmetric charge
neutralization; Cyclization;

Electrostatics; Folding; Protein-
induced DNA bending
Bending polarons, implication, 35
Benzo[a]pyrene diol epoxide (BPDE)
binding to DNA, 197–199
enantiomers, 197*f*
thermodynamic parameters of duplex
formation, 198*t*
B-form deoxyribonucleic acid (DNA)
structural characteristics in model,
139
structural deformations, 32–33
See also Deoxyribonucleic acid
(DNA); Dickerson–Drew
dodecamer (DDD)
Bifurcated hydrogen bonds, A-tract
curvature, 80, 83
Binding
endonuclease I to DNA junction, 156
structure selective, by junction-
resolving enzymes, 155–156
Bioinformatics, structural mechanism,
45, 48
Branch migration, DNA junctions,
154–155
Branch site helix
exposure of 2'OH nucleophile, 170
functional role of pseudouridine in
branch site, 172–173
pseudouridine-induced backbone
deformation, 169–170
schematics of unmodified and
pseudouridine-modified, 170*f*
structure of branch site duplexes,
167–169
water-pseudouridine interactions in
precursor messenger (pre-m)RNA,
171–172
See also Pseudouridine (Ψ)

C

Calorimetry
A-tract deoxyribonucleic acid
(DNA), 33

250

tRNAs, 178*t*
stacked X-structure of four-way
DNA junction, 146–149
Structure-selective binding, junction-
resolving enzymes, 155–156
Sugar puckering, A to B DNA
transition, 32
Superposition root mean square (rms)
dispersion, A-tract structures, 71,
74*t*

T

Temperature, duplex conformation at
high, 194
Thermodynamics
duplex containing benzo[a]pyrene
diol epoxide (BPDE), 198*t*
duplex formation containing
cisplatin, 200*t*
duplex formation of aminopropyl
cationic chains into Dickerson–
Drew dodecamer, 203*t*
netropsin binding to DNA, 195*t*
See also Folding
Transfer ribonucleic acid (tRNA)
dimerization in tRNA^{Leu} mutant, 184*f*
disease-related tRNA mutations, 178
effect of mutations producing CA
mispairs in hs mt (human
mitochondrial) tRNA^{Ile}, 181*f*
effect of mutations producing CA
mispairs in hs mt tRNA^{Leu}, 184*f*
hs mt genome, 178
impact of disease-related mutations
on structure and function of hs mt
tRNAs, 180
isoleucine in hs mt tRNA^{Ile} 180–183
leucine in hs mt tRNA^{Leu}, 183, 185
protein synthesis, 177
structural features and pathogenic
mutations in hs mt tRNAs, 178*t*
structurally weak TΨC stem within
hs mt tRNA, 181–183

unique structures of mitochondrial
tRNAs, 179
Transversions, AT to TA in middle of
A-tract, 82
TΨC stem, weak, human
mitochondrial transfer ribonucleic
acid (hs mt tRNA), 181–183

U

Uphelix structures
A-tract deoxyribonucleic acid
(DNA), 71, 72*f*
base inclination within A-tracts, 76*f*
normal vector plots of A-tract DNA,
73*f*
superposition rms distances, 74*t*
See also A-tract deoxyribonucleic
acid (DNA)
Uridine
structure, 166*f*
See also Pseudouridine (Ψ)
UVRR spectroscopy
A-tract DNA, 33
DNA sequences, 31*f*

V

Volume change
association reactions, 191–192
netropsin binding to DNA, 194–196

W

Water
conformational stability of nucleic
acids, 191–192
lower uptake by modified duplex,
204–205
pseudouridine interactions in pre-
mRNA branch site helix, 171–172
Water activity, structural
deformations, 32